Deploying Wireless LANs

Deploying Wireless LANs

Concepts, Operation, and Utilization

Gilbert Held

McGraw-Hill
New York • Chicago • San Francisco • Lisbon
London • Madrid • Mexico City • Milan • New Delhi
San Juan • Seoul • Singapore • Sydney • Toronto

McGraw-Hill

A Division of The McGraw·Hill Companies

1 2 3 4 5 6 7 8 9 0   AGM/AGM   0 9 8 7 6 5 4 3 2 1

ISBN 0-07-138089-2

The sponsoring editor for this book was Marjorie Spencer, the editing supervisor was Steven Melvin, and the production supervisor was Pamela A. Pelton. It was set in Vendome by Patricia Wallenburg.

Printed and bound by Quebecor/Martinsburg.

McGraw-Hill books are available at special quantity discounts to use as premiums and sales promotions, or for use in corporate training programs. For more information, please write to the Director of Special Sales, Professional Publishing, McGraw-Hill, Two Penn Plaza, New York, NY 10121-2298. Or contact your local bookstore.

 This book is printed on recycled, acid-free paper containing a minimum of 50 percent recycled, de-inked fiber.

To the students of Georgia College and State University.
The ability to learn from students while teaching is indeed
a valuable lesson.

▬▬ ▬▬ CONTENTS

Preface   xv

Acknowledgments   xvii

1   Introduction to Wireless LANs   1

Overview   3

    Operation   3

    Network Configurations   4

    Roaming   6

    Extension Points   6

Communications Methods   8

    Infrared   8

    Microwave   9

    Radio Frequency   9

    Benefits   11

Utilization   11

    Hospital Use   12

    College Use   13

    Inventory Control   13

    Internet Access   14

    Training Centers   14

    Facilitating Networking   15

    Trade Show Use   15

    Benefits   16

Constraints   17

Book Preview   18

    Terminology and Technology   19

    Understanding Wireless LAN Modulation   19

    Understanding Wireless LAN Communications Systems   20

    Wireless LAN Hardware   20

IEEE Wireless LAN Standards   21

Installing a Wireless LAN   22

The Home RF Standard   22

The Future   22

2   Terminology and Technology   23

Basic Communications Concepts   24

Powers of Ten   25

Frequency   26

Wavelength   27

The Frequency Spectrum   29

Bandwidth   32

Power Measurements   32

Signal-to-Noise Ratio   37

Transmission Rate Constraints   41

Nyquist Relationship   42

Radio Frequency Spectrum Allocation   45

U.S. Spectrum Allocation   46

Applications   49

Other Transmission Impairments   50

Basic Wireless LAN Components   51

Path Loss   52

Multipath Propagation   56

Fading   57

Enhancing Signal Reception   58

3   Understanding Wireless LAN Modulation   61

Basic Modulation Methods   62

Rationale   62

Modulation Process   63

Amplitude Modulation   64

Frequency Modulation   65

Phase Modulation   67

Contents

Wireless LAN Modulation Methods   74
  DSSS Modulation   75
  Differential Binary Phase Shift Keying (DBPSK)   75
  Differential Quadrature Phase Shift Keying (DQPSK)   76
  Complementary Code Keying (CCK) QPSK   78
  Frequency Hopping Spread Spectrum
    (FHSS) Modulation   79
  Gaussian Frequency Shift Keying (GFSK)   79
  Orthogonal Frequency Division
    Multiplexing (OFDM) Modulation   80
  Quadrature Amplitude Modulation (QAM)   82

4 **Wireless LAN Communications Systems**   **85**

Spread Spectrum Communications   86
  Development Rationale   86
  General Operation   87
  Spread-Spectrum Methods   88
  Direct Sequence Spread Spectrum (DSSS)   89
Frequency Hopping Spread Spectrum   91
  Regulations   91
  Operational Parameters   93
  Packet Transmission Capability   94
  Hopping Modes   94
  Advantages of Use   95
Direct Sequence Spread Spectrum   97
  Regulation   97
  Operation   98
  Using the Chipping Code   99
  Bandwidth Spreading   100
  Advantages of Use   101
  Disadvantages   102
Coded Orthogonal Frequency Division Multiplexing   103
  Evolution   103

Overview   105
Operation   107
Scrambling and Coding   108
Advantages of Use   110
Disadvantages   110

5   Wireless LAN Hardware   113

Wireless Access Point   114
Evolution   115
Equipment Connection   115
Using a Single Access Point   117
Using Multiple Access Points   118
Wireless LAN Network Cards   123
Wireless Bridges   127
Wireless Router/Gateway   136

6   IEEE Wireless LAN Standards   145

The 802.11 Standards   146
Overview   147
Topology   148
Portals   152
The Physical Layer   152
Modulation   154
Frame Format   155
Hopping Sequence   156
Direct Sequence Spread Spectrum   157
Overview   157
Modulation   157
Frequency Allocation   158
Frame Format   158
Infrared   159
Modulation   160
Frame Format   161

Contents

The MAC Layer   162
   Basic Access Method   162
   Minimizing Collisions   163
   Interframe Spaces   166
   Collision Avoidance   167
   Frame Types   167
   RTS Frame   177
   CTS Frame   177
   ACK Frame   178
Operation   179
   Joining an Existing Cell   179
   Authentication and Association   180
   Roaming   180
The 802.11b Standard Extension   182
   Overview   183
   Operation   183
   Modulation   183
The IEEE 802.11a Standard Extension   190
   Overview   190
   Modulation   191
   Frame Format   191
   Operation   193

7 Installing a Wireless LAN   195

The SMC Networks Barricade Router   196
   Product Overview   196
   Features   197
   Site Location   198
   Wireless Positioning   198
   Connectivity Tradeoffs   199
   Using WINIPCFG   200
Software Setup   202
   Verifying Computer—Router Connectivity   203

Configuring the Router		204
Configuration Options		205
Wireless Settings		216
Return to WINIPCFG		217
The SMC Networks EZ Connect PC Card		218
Driver Installation		219
Configuration Utility		222
Agere Systems Orinoco PC Card		227
Installation		227
The Client Manager		232
Proof of the Pudding		234

8   The Home RF Standard   237

Overview		238
Versions		239
Network Architecture		240
Nodes		240
System Requirements		241
Technical Characteristics		241
FHSS Use		241
Power, Operating Rate, and Modulation		242
Device Support		242
Security		243
Data Compression		244
Home RF Operation		244
The Physical Layer		245
The MAC Layer		245
Frame Duration and Types		246
Frame Operations		247

9   The Future   251

FCC Part 15 Ruling		252
Overview		252

ISM Band Use 253
RF Interference 254
The IEEE 802.11g Standard 255
Backward Compatibility Issues 256
Area of Coverage Consideration 257
The IEEE 802.1x Standard 257
Overview 258
Operation 258
Great Expectations 260
AiroPeek, A Wireless Protocol Analyzer 260
Overview 261
Capturing Traffic 262
Protocol Summary 264
Packet Decoding 265

A Hardware Manufacturers **267**

Locating Wireless Equipment on eBay 271
Equipment Location 272
The Bidding Process 277

B Wireless LAN Economics **279**

Limited Client-Based Wireless LAN 280
Access Point/Router-Based Wireless LAN 282
Wired LAN Access 283

C Practical Communications Security **285**

Glossary **289**

Index **303**

PREFACE

A few years ago a popular TV commercial closed with the saying "the future is now." In the wonderful world of data communications the future has arrived in the form of wireless LANs. Once considered a niche technology that was expensive and limited with respect to its data transfer capability, wireless LANs are now reaching into every corner of our lives. If we travel we will more than likely encounter public areas in airports and hotel lobbies equipped with access points that enable us to surf the Web, check e-mail, and perform other activities using our laptop, notebook, or PDA equipped with a compatible wireless LAN adapter. If we check into a hotel, visit a sporting event, or register for a college course, we may also encounter persons using computers with a wireless LAN capability to access data from servers and mainframes by first connecting to an access point on a wired LAN behind the scenes.

The ability to transmit and receive data without having a wired connection frees us to locate computing equipment nearer to the area where it is useful. If you visit a modern big city hotel lobby you may encounter a reception area in the middle of an atrium. Behind the counter a hotel employee has a computer connected to the hotel LAN. However, instead of a wire connection that might require the lobby floor to be dug up, the connection occurs via a wireless LAN. Not only is the connection less expensive but the time required to place the computer into operation behind the reception area is probably a small fraction of the time that would be required to establish a wired connection.

In a university environment it becomes possible to set up registration stations in a gymnasium very rapidly without running cables and having to temporarily tape them to the floor to alleviate the potential of students, administrators, and faculty tripping.

Similarly, libraries can add and remove workstation connections to the Internet in tandem with special events they may hold.

In a home or apartment environment it is becoming common for a digital subscriber line (DSL) or cable modem to be installed to obtain a high-speed broadband access capability. One key non-technical problem associated with the use of DSL and cable modems is the fact that your existing telephone and cable outlets may not be located in close proximity to your computer. Another related problem is the fact that many homes and apartments have multiple computers. Rather than rewire twisted pair or coaxial cable you can save time, avoid drilling holes in walls, and possibly save some money by installing a wireless LAN.

Today a wireless LAN provides us with the ability to communicate from locations that were previously difficult or impossible to support via wires. In addition, they provide a significant degree of flexibility and allow us to respond to changing requirements in a timely manner. Thus, when you think about networking, you should also think about wireless networking. When you do you will realize that the future is now!

As a professional author I value reader feedback. Although I have attempted to provide practical information throughout this book, I am human. To err is to be human, so if I omitted an area you feel I should have covered, spent too much time on a topic or assumed reader knowledge where a fuller explanation was warranted, please contact me. You can reach me either through my publisher, whose address is included on the jacket of this book or you can contact me via e-mail at gil-held@yahoo.com.

Gilbert Held
Macon, GA

ACKNOWLEDGMENTS

Many years ago I learned to appreciate the effort behind the publication of a book. While the author's name is prominently displayed on the cover of the book, it may require a considerable amount of effort to locate those other people responsible for the book you are now reading. I feel I would be remiss if I did not acknowledge the efforts of many people who contributed to the publication of this book.

First and foremost, once again I wish to thank my wife Beverly for her patience and understanding while I worked long evenings and on weekends preparing the manuscript. However, unlike my first experience with LANs, which required cabling the upstairs of our home, the use of a wireless LAN was probably more welcomed. No longer did I have to caution visitors nor block off the upper portion of our home to pets that could get tangled in cable.

As an old-fashioned author who frequently travels around our globe, I learned long ago that no matter what type of electrical adapters I took on a trip there would be a high probability that the hotel outlets would not accept my adapters. Thus, I continue to draft my manuscripts the old-fashioned way, using pen and paper. While I do not have to worry about my laptop or notebook running out of energy, I need to depend on typists who can read my handwriting, especially those scrawls resulting from air turbulence at 25,000 feet. Once again I am indebted to Mrs. Linda Hayes and Mrs. Susan Corbitt for converting my handwritten manuscript and drawings into a professional manuscript.

Another important person in the production of a book is one's acquisition editor. As a buffer between the writer, who may have tunnel vision about a project, and the marketing department that must sell the book, the acquisition editor must guide the proposal development process to satisfy both. Once again I am indebted to Marjorie Spenser for her fine effort.

As a "hands on" author who writes about technology, it is a pleasure to be able to work with new products. Thus, I would be remiss if I did not acknowledge the cooperation and assistance of Angie Tucker at SMC for arranging for me to enjoy firsthand use of wireless LAN equipment from her company.

Last but certainly not least, I would like to thank everyone involved in the production of this book. Those unseen heroes include the persons who proof the manuscript and galley pages, the production crew that arranges for its publication, the cover designer, indexer, and probably a few other people who track schedules and ensure the finished product is available on schedule. To each and every one of you a sincere thank you!

Deploying Wireless LANs

Introduction to Wireless LANs

Welcome to the emerging world of wireless LANs. In this book we will examine wireless LANs in detail, learning how they operate, where they can be effectively deployed, and the problems we should be aware of prior to their deployment. In this introductory chapter we start with the basics, then move into more detail and cover additional specifics in later chapters.

We commence our examination of wireless LANs with a definition: Because wireless LANs operate as an adjunct to wired LANs, we compare and contrast their operational capabilities with those of their wired cousins. Once we have a basic understanding of what the term "wireless LAN" means, we will turn our attention to examining how wireless LANs can be used. In doing so we will note that the use of wireless LANs can cover a wide spectrum of applications, from college and university campus applications to business, commercial, and residential use. As we examine several wireless LAN applications we will also note the benefits that can be obtained from their deployment. Because every coin has two sides, we will also examine some of the key limitations associated with the technology, to obtain a balanced view.

We conclude this chapter with a preview of succeeding chapters. This preview can be used as is, or in conjunction with the table of contents and index to locate information of immediate concern. While this book was structured to provide readers with a logical flow of material with the objective of learning significant details concerning the operation of wireless LANs, I'm also a realist. From working with other professionals, teaching college courses, and lecturing I recognize that, on occasion, readers need to determine specific information concerning a topic and do not have sufficient time to read, for example, three chapters to understand a point of interest in the fourth. Recognizing this fact of life, where possible, each chapter was written to be as independent as possible of other chapters in this book. If you are new to communications or a bit rusty on terminology and technology, however, it is highly recommended that you read Chapter 2 prior to directly accessing information of immediate interest. That said, grab a Coke and a few munchies, and follow me into the world of wireless LANs.

Overview

We can define a wireless local area network (LAN) as a communications system that uses radio frequency (RF), microwave, or infrared to interconnect devices within a limited geographic area. A wireless LAN can be used as an extension to a wired LAN or independently as an alternative to a wired LAN. Thus, wireless LANs provide a great deal of networking flexibility. Now that we have a very basic description of what a wireless LAN represents let's build on our definition and examine how a wireless LAN operates.

Operation

Unlike wired LANs, where transmission occurs over a copper or fiber optic conductor, a wireless LAN uses the electromagnetic frequency spectrum to communicate information. Similar to radio and television, wireless LANs transmit information using airwaves. Transmission can occur using radio microwave or infrared, with the frequency available for use and the level of allowable transmit power commonly regulated by a governmental agency. In the United States the *Federal Communications Commission* (FCC) is the regulatory agency that controls the use of the frequency spectrum and the level of transmit power that can be used.

Although a wireless LAN uses the ether instead of a copper or fiber optic medium for transmission, the actual manner by which data is transmitted is similar to that of its wired cousin. That is, both wired and unwired LANs operate by converting binary 0s and 1s into signals suitable for the transmission media used. In wireless LANs, transmission primarily occurs by impressing information (*modulating*) on a radio signal referred to as a *carrier*. As we note later in this book, the most common types of wireless LANs borrow technology originally developed for military applications and use some techniques where a radio signal literally spreads out or hops among different frequencies within a band of frequencies allocated for the technology. A receiver in this type of wireless

LAN environment can be considered similar to a sophisticated radio, tuning itself to a specific frequency to receive some information and then to another frequency to continue its reception. The transmitter and receiver must obviously be synchronized, and later in this bookwe describe how that is accomplished.

Network Configurations

There are two basic network structures supported by wireless LANs. The first is relatively simple and involves installing two or more devices with wireless adapter cards to enable them to communicate with one another when they are within a certain distance. This type of network configuration is referred to as a peer-to-peer network.

PEER-TO-PEER NETWORKING. Although we might normally consider two PCs as representing a peer-to-peer wireless LAN, in actuality this is a minor use of this network configuration, because it is often quicker and less expensive to use a zip disk or a cable and a software program to exchange files between PCs. Instead, wireless peer-to-peer networking more commonly occurs in the home as one of several methods used to extend a single broadband digital subscriber line or cable modem connection.

Figure 1.1 illustrates a simple peer-to-peer wireless network. In this example, a small wireless LAN-compatible router was installed in a home between a broadband termination device and one hardwired computer. Assuming that the homeowner does not wish to use his crawl space or ceiling for wiring experimentation, the use of a peer-to-peer wireless network from the router to the second PC may eliminate drilling holes and running cable.

WIRELESS-TO-WIRED NETWORKING. A second wireless network structure is designed to support transmission from one or more wireless devices onto an existing wired LAN. Doing so permits wireless devices to gain access to the resources of the wired LAN.

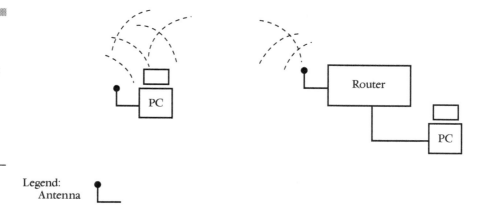

Figure 1.1
A peer-to-peer wireless network configuration occurs when two devices communicate with one another independent of a fixed network structure.

Legend:
Antenna

Figure 1.2 illustrates a wireless-to-wired network configuration. In this network configuration or structure one or more access points are required to serve as an interface between the wired LAN and the nodes in the wireless LAN. The access point typically serves between 10 and 40 nodes, keeping track of transmission and reception to and from each node, buffering received data for transmission onto the wired LAN and, in the reverse direction, buffering data received from the wired LAN that is destined for nodes on the wireless LAN.

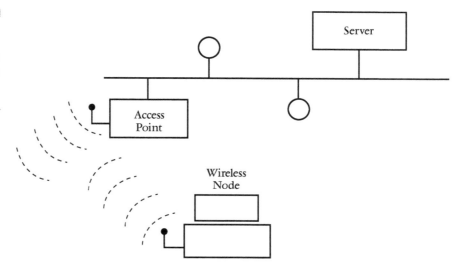

Figure 1.2
A wireless-to-wired network requires one or more access points to service wireless nodes.

An access point services a finite geographic area, commonly supporting transmission within a radius of 300 to 500 feet from the device. The exact distance supported depends on the power level of the signals generated by the access point and wireless nodes, obstructions within the geographic area to be served, and the presence or absence of other signal sources, which include electromagnetic radiation produced by other electronic devices within proximity of the wireless LAN. For a large building or campus, there may be several access points installed to provide connectivity between wireless nodes and the wired LAN.

Roaming

A person who moves from one location to another should be recognized as leaving one access point and joining another, a condition referred to as *roaming* and one which cellular phone users take for granted every day. In a roaming situation, one access point must recognize that a signal is becoming weaker and *hand-off* the client to the next access point. Figure 1.3 illustrates an example of two access points used to provide connectivity within a large building. In examining Figure 1.3, note that very rarely do computer users roam from one access point to another. Instead, they primarily set up a laptop or notebook at a location that would otherwise not be accessible and work from that fixed location. Certain types of devices, however, including hand-held PDAs, incorporate wireless LAN support and require the ability to roam. As we note later in this chapter when we turn our attention to wireless LAN applications, many retail organizations use PDAs and specialized notebooks in a roaming environment.

Extension Points

If you are familiar with the manner in which wired LANs operate, you are probably familiar with the term *repeater*. A repeater looks for a digital pulse and regenerates the pulse, permitting a

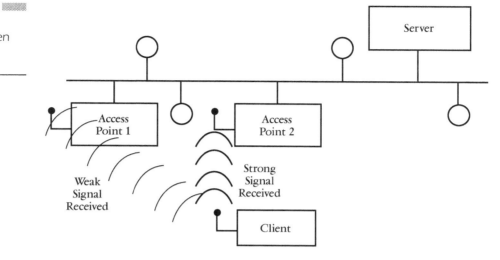

Figure 1.3
Roaming between
multiple access
points.

As the client moves away from access point 1 towards access point 2, the strong signal received by the second access point becomes weaker while the signal received by the first access point becomes stronger.

Lengend:
    Antenna

signal to flow a further distance down a transmission path. In a wireless LAN environment a special type of repeater is referred to as an *extension point*.

An extension point, as its name implies, extends the transmission range of a wireless LAN. To accomplish this task the extension point receives a signal from a node and remodulates the signal. Thus, an extension point can be viewed as operating much as a radio relay station does.

Figure 1.4 illustrates the use of an extension point to extend the range of a wireless LAN. While you can consider installing a series of extension points to extend the coverage of a wireless LAN from a particular access point to a remote node, doing so is similar to the "Christmas bulb effect." If one extension point in a series of extension points should fail, the ability of a remote mode to access the wired LAN will be terminated. In addition, each extension point requires power and shelter, which can rapidly increase the cost associated with adding a series of extension points to extend access to a wired LAN from one or more remote locations. Because of these drawbacks, an extension point is commonly used on an individual basis.

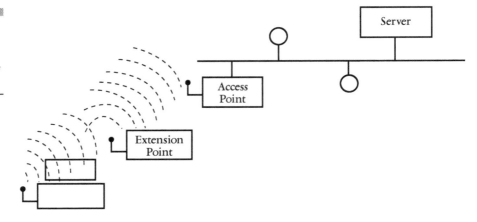

Communications Methods

Wireless LANs employ one of three transmission methods, using infrared, radio frequency, or microwave.

Infrared

Infrared (IR)-based wireless LANs are simple to design and, as a result, are relatively inexpensive. IR systems use very high frequencies (just below visible light in the electromagnetic spectrum) as the mechanism for transmitting data. IR systems are also relatively easy to design, because they need only to detect the amplitude of a signal. In addition, unlike radio frequency and microwave-based systems, infrared is not regulated by the FCC. Unfortunately, infrared cannot penetrate opaque objects. This means that walls, dividers, coat racks, and even persons walking across a room can obstruct an IR signal.

The most common form of an IR-based wireless LAN uses *line-of-sight* transmission. I remember seeing an early IR-based wireless LAN installed during the late 1980s where transceivers mounted on walls were pointed to other transceivers. Obviously this was not conducive to mobility.

A second type of IR-based wireless LAN is a *diffuse* or *reflective* wireless LAN. Although this type of system does not require a line-of-sight path from transmitter to receiver, its range of coverage is limited to a single room.

Microwave

Microwave transmission systems are limited by the FCC to 500 milliwatts of power. Such systems modulate data at a single frequency and primarily operate in the 5.8 GHz frequency band. This band, which extends from 5.725 through 5.85 GHz, represents one of several frequency bands allocated to industrial, scientific, and medical (ISM) operations without requiring a license. Thus, the 5.8 GHz band represents an unlicensed band. Although the use of the 5.8 GHz band provides a high throughput, it represents a nonstandard approach to wireless LANs and has only a minor share of the market for wireless LAN products.

Radio Frequency

By far the most popular types of wireless LAN products are based on the use of *radio frequency* (RF). RF-based wireless LAN products must operate in the 2.4 to 2.4853 GHz and 5 GHz frequency bands. While both bands are unlicensed, the FCC places restrictions on ISM bands. In the United States RF-based wireless LANs have certain power restrictions depending on the band they operate in. In addition, RF-based wireless LANs must use *spread spectrum technology,* and the FCC defines the general manner by which such technology can operate in each band.

There are three types of spread spectrum used with wireless LANs—*frequency hopping spread spectrum* (FHSS), *direct sequence spread spectrum* (DSSS), and *orthogonal frequency division multiplexing* (OFDM). The first two techniques owe their origin to the military, which required a communications technology that would be hard to jam.

FREQUENCY HOPPING SPREAD SPECTRUM. Frequency hopping spread spectrum (FHSS) splits the available frequency band into a series of small subchannels. A transmitter hops from subchannel to subchannel, transmitting short bursts of data on each channel for a predefined period, referred to as *dwell time*. The hopping sequence is obviously synchronized between transmitter and receiver to enable communications to occur. FCC regulations define the size of the frequency band, the number of channels that can be used, and the dwell time and power level of the transmitter.

DIRECT SEQUENCE SPREAD SPECTRUM. Direct sequence spread spectrum (DSSS) represents a second type of spread spectrum technology employed by RF-based wireless LANs. Using DSSS, the transmission signal is spread over an allocated frequency band that is also regulated by the FCC in the United States. Under DSSS a pseudo random binary string is used to modulate the signal. This string is referred to as a *spreading code,* and it is used to map data bits into a larger number of bits that provide redundancy. The mapping of data bits into a pattern is called a *chip* or *chipping code.* The number of chips used to represent a bit is known as the *spreading ratio.* As you might surmise, both transmitter and receiver must use the same spreading code, and the size of the spreading ratio is also regulated by the FCC.

ORTHOGONAL FREQUENCY DIVISION MULTIPLEXING. Orthogonal frequency division multiplexing (OFDM) represents one of the most popular methods used to obtain high speed communications in a broadband environment. Using OFDM multiple carriers, referred to as *subcarriers,* are used orthogonally or independently of each other. Data encoded onto each subcarrier cumulatively results in a high rate of data transport. One of the key advantages obtained from the use of OFDM is the fact that the separation of data onto a series of subcarriers separated from one another results in less intersymbol interference than when a single carrier is used. In Chapter 2 we discuss intersymbol interference, and in Chapter 4 we discuss wireless communications systems using FHSS, DSSS, and OFDM in greater detail.

Benefits

Both FHSS and DSSS represent transmission methods that minimize the effect of impairments to system throughput. Using FHSS, hopping from frequency to frequency minimizes the effect of electromagnetic radiation. Using DSSS, the spreading of data bits based on spreading codes can also serve as a method to minimize transmission errors (which we note later in this book). OFDM limits intersymbol interference by distributing modulated data over multiple subchannels.

Now that we have a general appreciation for the manner in which wireless LANs operate, let us turn our attention to how they can be used and the benefits they provide.

Utilization

One of the best ways to appreciate the benefits of a technology is to discuss its potential utilization. Thus, let us focus our attention on some possible areas where wireless LANs can be used. Table 1.1 lists seven generic application areas we examine in this section.

TABLE 1.1

Examples of
Wireless LAN
Application Areas

* Recording and delivery of patient information within a hospital
* Facilitating special events on a college campus
* Controlling inventory for wholesale and retail applications
* Accessing the Internet from portals in hotels, airports, and public buildings
* Setting up ad-hoc training centers configured on short notice
* Eliminating adds, moves, and changes in a dynamic network environment
* Facilitating trade show operations

Hospital Use

When I worked for a computer manufacturer approximately 30 years ago, one of the hot buzzwords at that time was "*clinical laboratory*." In the minicomputer section of a computer manufacturer during the 1970s, hospitals were viewed as a needy target for automation. While many competitors targeted back office operations, the company I worked for at that time was rather bold in the fact that they focused their attention on attempting to integrate patient room electronics to the hospital's X-ray and laboratory, calling the system a clinical laboratory. Unfortunately, the clinical laboratory system was probably a bit before its time, with the need to place relatively large and noisy terminal devices in patient rooms a major problem only solved by advances in the evolution of laptop, notebook, and PDAs. Today the use of wireless LANs in hospitals provides doctors and nurses with rapid access to patient data, including various laboratory reports. In addition, the input of data removes the problems many hospital workers have when attempting to decipher handwritten notes.

Wireless LANs in a hospital environment also excel in pill dispensing. As a nurse prepares to move from room to room, he can use a small notebook or a PDA with a wireless LAN card to communicate with a server on the hospital wired LAN, verifying the types and dosage of pills to be placed in the cart. As the nurse moves from room to room, the notebook or PDA generates appropriate screen displays that serve as an additional safeguard to verify that the correct pills and prescribed dosage is given to patients.

College Use

In a college environment a wireless LAN provides the flexibility to support special events in an economical manner. For example, consider school registration, which may occur two or three times a year. This is usually held in the gymnasium where it is difficult to extend a wired LAN that requires cables to be taped to the floor; the wireless LAN permits registration stations to be located where they make sense and not at a location that is simply convenient for cabling. Similarly, other college activities occurring on a periodic basis, such as alumni events and sporting events that may require the use of computational facilities at locations distant from a wired LAN, are also candidates that can benefit from the mobility and rapid installation afforded by the use of a wireless LAN.

Inventory Control

Many years ago when I had insomnia and took a ride to a drug store, I encountered store employees walking down the aisles with a tape recorder, speaking into the recorder as they counted items on a shelf. Being curious, I naturally asked an employee what he was doing. Believe it or not, this was my first introduction to inventory control.

If we fast forward to our modern era we might encounter another employee walking the aisles. However, instead of talking into a tape recorder whose tape would be mailed to a corporate office for transcription, the employee now walks the aisle with a different device. That device may be a special type of PDA with a bar code reader, enabling the employee to scan items on the shelves. Periodically, the employee can press a transmit key and send the contents of the PDA to an access point in the store that enables scanned inventory to reach the store LAN. From there the scanned data may flow via a terrestrial or satellite link to a corporate data center.

Internet Access

One of the more interesting features being advertised by airlines and hotels is the ability for consumers to obtain high speed Internet access when they use facilities at certain locations. Some hotels now offer Internet access via wireless LANs both from public areas and conference rooms that are referred to as *Internet portals.* Visitors can either use their own computers with an appropriate wireless LAN adapter card, rent a card from the hotel for use in their laptop or notebook, or rent a PC that is wireless LAN capable.

Another common location for Internet portals is airports. Both airlines and private organizations are offering Internet access not only at many airports in the United States but throughout major airports around the globe.

Training Centers

Corporate training is a multibillion dollar industry. Corporate training includes attendance at seminars offered at numerous locations in most major cities as well as in-house training when a sufficient number of employees makes it economical to offer a course on site. For both onsite and offsite training many times computer access is required. Once again, the use of a wireless LAN can provide an organization with the ability to rapidly structure a classroom and provide networking capability without having to worry about cabling.

Facilitating Networking

One common benefit is that a wireless LAN facilitates dynamic networking. Need workstations in a parking lot to gain access to the wired LAN? Use a wireless LAN. Need Internet access in a hotel lobby? Use the hotel wireless LAN portal. Need to quickly distribute the applicable pills and dosage to patients? Use a wireless LAN. Thus, it should come as no surprise that in a business, government, or academic environment where employees move about from one location to another, wireless LAN will make the life of a network manager or LAN administrator much simpler. Instead of having to worry about plugging and unplugging hub connections and installing wire to new areas, the network manager or LAN administrator can install access points at applicable locations throughout the organization. Then, when an employee arrives at work on a Monday morning and receives an assignment to temporarily relocate to marketing, that person only needs to move his or her PC to the new location.

Trade Show Use

In concluding our brief tour of wireless LAN applications let us turn our attention to the ubiquitous trade show. While it is fun to attend Comdex, Interop, or ComNet, it is more than likely anything but fun for the employees of firms who have to set up the booth for their organization. In the past, demonstrations had to be carefully planned and the cost of installing cable more often than not could be quite expensive because of union rules that required several "engineers" to work a minimum of a few hours to snake a thirty-foot cable through a conduit. Today many organizations have turned to the use of wireless LANs. Not only does the use of a wireless LAN alleviate the delay and cost of installing temporary wiring but the booth manager now controls the booth. Thus, a strike of the electrical brotherhood or a no-show of the authorized cable installer now has a minimal effect on the organization's booth.

Now that we have an appreciation for seven generic application areas where wireless LANs can be effective let us turn our attention to the benefits they afford.

Benefits

In discussing seven generic wireless LAN applications we noted several benefits obtained from their use. First, they are easy to use and provide flexibility in responding to organizational requirements. Second, they reduce the need for conventional cabling that either may not be possible or may be expensive. Thus, unless a portal operator charges an exorbitant fee for Internet access a wireless LAN can reduce the overall cost of operation for both operator and user. Third, wireless LANs obviously provide mobility and facilitate adds, moves, and changes. Another benefit only peripherally mentioned is scalability. By either placing or adding access points and extension points at appropriate locations, it becomes possible to satisfy expanding organizational requirements. Table 1.2 summarizes benefits that may be achievable through the use of a wireless LAN.

TABLE 1.2

Potential Benefits of Wireless LANs

Ease of use
Flexibility in meeting organization requirements
Can reduce overall cost of operation
Provide user mobility
Facilitate adds, moves, and changes
Scalability

Now that we have an appreciation for the benefits associated with the use of a wireless LAN it is important to note that such LANs are very beneficial but not perfect. Let us turn our attention to their constraints.

Constraints

There are several areas associated with the use of a wireless LAN that function as constraints on the ability to transmit and receive data. Five of the major constraints associated with the use of wireless LANs are listed in Table 1.3.

TABLE 1.3

Wireless LAN Constraints

Transmission range

Throughput

Interference

Cost

Battery life of mobile platforms

The low power and high frequency of wireless LAN devices limits their transmission range. Whereas conventional wired LANs can have a range on the order of kilometers through the use of fiber optic repeaters, wireless devices have a range in only hundreds of meters.

Until the turn of the century the maximum transmission rate of wireless LANs was 2 Mbps. The introduction of equipment supporting the IEEE 802.11b standard increased throughput to 11 Mbps; once equipment that complies with the IEEE 802.11a standard reaches the market, throughput up to 54 Mbps may be possible.

While older technology may appear to be a bottleneck when compared to the operating rates of wired LANs, it is probably more important to consider the number of nodes contending for an access point rather than raw throughput. For example, assume you are comparing a 802.11b wireless LAN to a Fast Ethernet LAN operating at 100Mbps. Suppose you plan to link a wireless LAN to a previously unpopulated 100BaseT segment via a single access point to service five nodes. Let us also assume that you have eighty nodes on the Fast Ethernet LAN.

In comparing the wireless LAN to the wired LAN we can divide the operating rate by the number of nodes to obtain a comparison of data rate per node on each type of LAN. For the wired LAN 100Mbps/80 results in an average rate of 1.25Mbps per node. For the wireless LAN note that even though the access point is connected to a wired LAN operating at 100Mbps, the access point for an 802.11b LAN is limited to supporting a data rate of 11Mbps. Thus, 11Mbps/15 results in an average data rate of .733Mbps per node. Note from the preceding that one method you can use to enhance the transmission capacity of wireless LANs is to install access points on unpopulated wired segments. A second method is to adjust the number of nodes per access point to satisfy organizational requirements.

Returning to our discussion of wireless LAN constraints, interference caused by multipath propagation can be a constraint on throughput. Similarly, electromagnetic interference (EMF) can adversely effect transmission. Thus, an appropriate site survey may be able to nip many problems in the bud.

Until a few years ago wireless LAN adapter cards and access parts were relatively expensive. Although the cost of both products has been reduced because of economics of scale from increased production, it is still many times that of 10BaseT network cards.

The last major wireless LAN constraint we discuss is the battery life of the mobile platform. When a wireless LAN is used to provide communications from a location where it is difficult to run a cable to an existing LAN, it is more than likely that that location lacks electrical outlets. Similarly, as a person using a PDA moves through a store taking inventory, the availability of outlets is not meaningful because it takes time to charge the device's battery. Thus, in many situations the battery life of a mobile platform can be a considerable constraint that be considered.

Book Preview

In this section we preview succeeding chapters in this book to provide you with a guide to the material to be presented and to supplement the table of contents and index when you want to locate information of interest. Although this book was written to be read in sequence, many of you need instant access to specific material and do not have the luxury of time to read a book from cover to cover. Recognizing this fact, I have structured material in each chapter to be as independent as possible from preceding and succeeding chapters. While this book structure supports immediate access to specific information, if you are new to the field of communications or are not conversant in certain technical areas, reading Chapter 2 prior to reading another chapter is highly recommended. This said, let us again grad our favorite beverage and a few munchies and take a tour of this book.

Terminology and Technology

In Chapter 1, we define wireless LANs and explore their benefits, uses, and limitations. In Chapter 2 we begin our technical investigation of wireless communications, learning or (for some readers) reviewing the concepts of frequency, bandwidth, and wavelength, understanding of the difference between bit and baud, and discovering how these terms relate to two basic equations that govern the transmission capacity obtainable on a medium. Continuing our investigation of basic terminology and technology associated with wireless LANs, we examine the frequency spectrum, note where wireless LAN transmission occurs, and become knowledgeable concerning different transmission impairments that affect wireless transmission. Because we need to know how to measure the level of signal power received to understand the effect of impairments, we also examine the role of the Bel and decibel in this chapter.

Understanding Wireless LAN Modulation

In Chapter 3 we examine how wireless transmission occurs. We get an appreciation for the rationale for modulation and the basic modulation methods used to convey data over a transmission medium. Once this is accomplished, we examine combined modulation methods, such as the use of amplitude and phase shift. This provides us with a foundation for examining how spread spectrum RF systems operate.

Understanding Wireless LAN Communications Systems

We can define a communications system as representing the method by which data is transmitted within a designated frequency band. In a wireless LAN environment there are three main types of communications systems used to transmit data, and we cover them all in Chapter 4. These communications systems include frequency hopping spread spectrum (FHSS), direct sequence spread spectrum (DSSS), and orthogonal frequency division multiplexing (OFDM).

Wireless LAN Hardware

The first four chapters of this book represent technical background material. In Chapter 5, we start to put the pieces of wireless LAN hardware together by examining in detail some basic components. In this chapter we examine wireless LAN cards, access points, bridges, and routers. In doing so we examine how such equipment operates, the use of software required to make such equipment interoperable, and several topology examples that illustrate the use of different hardware in a wireless environment. It should be noted that Chapter 5 provides generic information and serves as an introduction to the more detailed information presented n the next chapter, which is focused on three IEEE wireless LAN standards.

IEEE Wireless LAN Standards

Chapter 6 provides you with detailed information concerning three Institute of Electrical and Electronic Engineering (IEEE) standards, referred to as the IEEE 802.11, 802.11a, and 802.11b standards. The original IEEE 802.11 standard was ratified in 1997 and defined three types of wireless LANs that could operate at either 1 or 2 Mbps. In September 1999 the IEEE approved an extension to the 802.11 standard. Designated as 802.11b, this standard upped the data rate of wireless LANs to 5.5 Mbps and 11 Mbps while providing backward compatibility with one of the three types of LANs defined under the original specification. Both the original IEEE 802.11 standard and the 802.11b extension operate in the 2.4 GHz unlicensed industrial, scientific and medical (ISM) frequency band. The second extension to the 802.11 standard operates in the 5GHz frequency band and is designated as the 802.11a standard. In Chapter 6 we first focus our attention on the 802 and 802.11b standards that operate in the 2.4 GHz frequency band. Once this is accomplished we turn our attention to the 802.11a standard that operates in the 5 GHz frequency band. For each of the three IEEE standards we examine the basic components of the standard, the structure of frames and their channel assignments, and other technical information associated with each standard.

If you are a bit puzzled as to the placement of this chapter after instead of before LAN hardware, let me explain: Wireless LAN hardware is applicable to all three IEEE standards, however, at the present time the IEEE 802.11a standard remains to be implemented by manufacturers. In addition, basic wireless LAN hardware is applicable to other wireless LAN standards. Thus, in Chapter 5 we examine hardware as independently as possible from different standards, only noting and referencing standards when applicable. This enables us to examine each of the three IEEE 802.11 standards as an entity in Chapter 6. After we obtain an appreciation for both IEEE standards, Chapter 6 focuses on installing an 802.11b LAN.

Installing a Wireless LAN

In Chapter 7 we turn to specifics, following my adventure in installing and operating a wireless LAN. In this chapter I relate my experiences concerning topology considerations, deployment methods, and techniques I used to enhance transmission.

The Home RF Standard

While both IEEE 802.11 standards are oriented towards business, government, and commercial use, a second standard, referred to as the *Home RF standard,* is oriented towards home users. In Chapter 8 we examine this standard and compare and contrast its products to IEEE 802.11—compatible products.

The Future

In Chapter 9 I polish my crystal ball and peer into the future. Where wireless LANs are headed and what new applications are on the horizon are just two items of concern that we examine in Chapter 9. Now that we have an appreciation for where we are going, let us begin our exploration of wireless LANs by defining a wireless LAN and examining the basic technology.

Terminology
and Technology

In this chapter, we'll become acquainted with the frequency spectrum, its use, and some basic relationships that govern the operation and utilization of wireless local area network (LAN) systems. Because readers have different backgrounds, we commence with a review of basic technology associated with communications, examining what is meant by such terms as frequency, bandwidth, and wavelength. Because the focus of this book is on the transmission of data over different types of wireless LAN transmission systems, we also examine the relationship between bandwidth and the theoretical obtainable data transmission rate. This examination includes a discussion of bits and bauds as well as a pair of classic relationships and laws governing the obtainable data rate on a channel—the Nyquist Relationship and Shannon's Law. After this is accomplished we will discuss the management of the frequency spectrum and present some examples of the allocation of different frequency bands to different applications. In doing so we note the allocation of a portion of the frequency spectrum for use by different types of wireless LANs. Because one of two bands in the allocated spectrum is also available for use by other applications, we discuss some of the problems that can occur when sharing a portion of the frequency spectrum. We will conclude this chapter by noting some additional transmission impairments that can adversely affect the ability to use a wireless LAN. These impairments include intersymbol interference, propagation delay, attenuation, path loss, and the effect of signals taking numerous paths to a receiver because of reflections referred to as multipath transmission. Let's start with the basics, first reviewing the prefixes for the powers of ten since their use is commonly employed in any discussion of wireless communications.

Basic Communications Concepts

Because readers have diverse backgrounds, some of us may be more conversant than others with acronyms used to reference powers of ten, what the term frequency means, the relationship between wavelength and frequency, and similar topics that provide a foundation of knowledge necessary to understand how wireless LANs operate. In this section we focus our attention on basic communications concepts to ensure that we all have an equivalent background of communications knowledge.

Powers of Ten

As a refresher for those of us who may be a bit rusty remembering the prefixes for the powers of ten, Table 2.1 lists seven common prefixes and their meaning. In the wonderful world of wireless communication we commonly encounter the terms MHz and GHz when discussing the operating frequency of different communications systems. Because frequency and wavelength are inversely related to one another, as frequency increases wavelength decreases. Thus, while we use relatively large powers of ten for frequency, we use relatively small powers of ten to denote wavelength.

TABLE 2.1

Common Prefixes
of Powers of Ten

Prefix	Meaning	
nano-	1/1,000,000,000	billionth
macro-	1/1,000,000	millionth
milli-	1/1000	thousandth
kilo-	1000	thousand
mega-	1,000,000	million
giga-	1,000,000,000	billion
penta-	1,000,000,000,000	trillion

Frequency

Frequency refers to the number of periodic oscillations or waves that occur per unit time. Figure 2.1 illustrates two oscillating sine waves at different frequencies. The top portion of Figure 2.1 illustrates a sine wave operating at one *cycle per second* (cps). Note that the term cps has in general been replaced by the synonymous term *Hertz*, abbreviated as Hz and named in honor of the German physicist. The lower portion of Figure 2.1 shows the same sine wave after its oscillating rate is doubled to 2 Hz.

Figure 2.1
Oscillating sine waves at different frequencies.

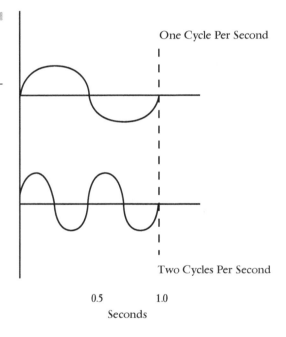

The time required for a signal to be transmitted over a distance of one wavelength is referred to as the *period* of the signal. That period represents the duration of the cycle and can be expressed as a function of the frequency. That is, if *T* denotes the period of the signal and *f* is its frequency, then:

$$T = 1/f$$

We can also express frequency in terms of the period of a signal. Doing so we obtain:

$$F = 1/T$$

Based on the preceding, we note that the sine wave shown in Figure 2.1, whose signal period is 1 second, has a frequency of 1/1 or 1 Hz. Similarly, the second sine wave, whose period is 0.5 seconds, has a frequency of 1/0.5 or 2 Hz. Thus, as the period of a signal decreases, its frequency increases.

Wavelength

The period of an oscillating signal is also referred to as its *wavelength* (λ). We can obtain the wavelength in meters by dividing the speed of light in free space (3×10^8 meters/s) by the frequency in Hz. That is:

$$\lambda_{meter} = \frac{3 \times 10^8}{f_{Hz}}$$

Note that we can also adjust the numerator and denominator of the preceding equation. Doing so we can adjust the frequency from terms of Hz to KHz, MHz, and GHz to determine the wavelength as follows:

$$\lambda_{meter} = \frac{3 \times 10^8}{f_{Hz}} \quad \frac{3 \times 10^5}{f_{KHz}} \quad \frac{300}{f_{MHz}} \quad \frac{0.3}{f_{GHz}}$$

Because the wavelength is expressed in terms of the speed of light divided by the frequency, we can also define the frequency in terms of the speed of light divided by the wavelength. Doing so we obtain:

$$f_{Hz} = \frac{3 \times 10^8}{\lambda_{meter}}$$

Just as we can compute the wavelength in terms of varying frequency, we can compute frequency in terms of varying the speed of light constant. That is, we can compute frequency in terms of a wavelength in meters as follows:

$$F_{hz} = \frac{3 \times 10^8}{\lambda_{meter}} \qquad f_{KHz} = \frac{3 \times 10^5}{\lambda_{meter}} \qquad f_{MHz} = \frac{300}{\lambda_{meter}} \qquad f_{GHZ} = \frac{0.3}{\lambda_{meter}}$$

Although we normally like to be precise with our computations there are several rules of thumb that can be used to expedite computation. For metric computations you can estimate the wavelength in centimeters as follows:

$$\lambda cm = 30/f_{GHz}$$

For example, assume a wireless system operates at 10 GHz. Then, its wavelength is approximately 30/10 or 3 cm.

For English measurements we can estimate the wavelength in feet as follows:

$$\lambda ft = 1/fGHz$$

Returning to our previous example, where the frequency is 10 GHz, the wavelength then becomes 1/10 or 0.1 feet.

The wavelength of a signal has a significant impact on the size of an antenna. For example, at low frequencies the wavelength becomes quite long. This explains why U.S. Navy submarines that use a low frequency transmission system to communicate when submerged eject an antenna (in the form of a length of wire) that can be thousands of feet in length. This also explains why it takes a long time to transmit messages when submerged and why the U.S. Navy restricts the length of personal messages from family members to seamen. Concerning the latter, we impress information on an oscillating signal by varying its amplitude, frequency, or phase. Because bits are converted into signal changes referred to as bauds, a low frequency transmits at a lower data rate than does a higher rate of signal change. Thus, transmission at a low frequency requires more time to send a signal than does a higher frequency. In comparison, wireless LANs that operate in the GHz portion of the frequency spectrum have a relatively short wavelength, which enables wireless telephones to be fabricated with relatively short antennas. Because wireless LANs operate in the high frequency spectrum there are more signal changes per unit time. In general, this enables more data bits to be transmitted per unit time than do lower operating frequencies.

The Frequency Spectrum

The frequency spectrum ranges from 1 Hz to at least 10^{23} Hz, with the possibility that there are frequencies beyond cosmic rays waiting to be discovered. Figure 2.2 provides a summary of the currently known electromagnetic radiation spectrum, with the placement of several common communications applications indicated at appropriate points on the chart. In examining Figure 2.2, note that the right column indicates the wavelength associated with a corresponding frequency. As we would expect from our preceding examination of the relationship between frequency and wavelength, we note from the chart that this relationship is inverse.

Figure 2.2

Electromagnetic
radiation spectrum.

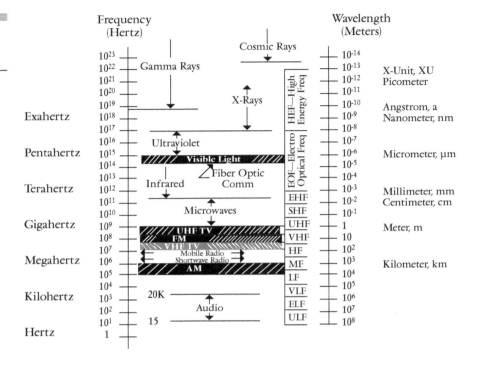

In examining Figure 2.2, note the frequency in the GHz (10^9 or billions of Hz per second) range. The Federal Communications Commission (FCC), which regulates the use of the frequency spectrum in the United States, allocated the 2.4 GHz spectrum to industrial, scientific, and medical (ISM) applications. Applications do not require licenses to operate in the 2.4 GHz band; however, they are limited with respect to their operating power. Because of this, the 2.4 GHz frequency spectrum is also referred to as the *unlicensed ISM band*.

In addition to the 2.4 GHz band's being unlicensed in the United States, this frequency band is also unlicensed on a global basis; however, the exact frequencies and power can vary from country to country. Table 2.2 lists the frequency range and bandwidth of the 2.4 GHz ISM band for the United States, most of Western Europe, Spain, France, and Japan. Note that the term *bandwidth*, which we will shortly cover in more detail, denotes the width of a range of frequencies. The first generation of wireless LANs operated in the 2.4 GHz ISM band. Other wireless transmissions, including Bluetooth-compatible devices, the Home RF standard, and microwave ovens, also operate in the ISM band and interference can occur between different applications.

A second generation of wireless LANs, which was standardized in 1999 and for which products are expected to reach the market during 2002, is designed to operate in the 5 GHz transmission band. Not only will such equipment operate outside the interference of 2.4 GHz products but this equipment will be able to use more bandwidth that supports higher data transmission rates than currently obtainable. Later, we describe and discuss the operation of both generations of wireless LANs.

TABLE 2.2

The 2.4 GHz ISM Band Can Vary by Location

Location	Frequency Range (GHz)	Bandwidth (MHz)
United States	2.4000—2.4835	83.5
Western Europe*	2.4000—2.4835	83.5
Spain	2.4450—2.4750	30.0
France	2.4465—2.4835	37.0
Japan	2.4710—2.4970	26.0

*For most countries in Western Europe.

Returning to Figure 2.2, also note that fiber optic communications is relatively high on the frequency portion of the chart. Most of us are probably aware of the fact that many recently formed communications carriers based their infrastructure on the construction of tens of thousands of route miles of fiber cable because the range of frequencies supported—referred to as bandwidth—is extremely high. Because the transmission rate is proportional to bandwidth, this allows one fiber to replace thousands of copper cables. Now let us turn our attention to bandwidth and how it governs transmission capacity.

Bandwidth

Bandwidth is the measurement of the width of a range of frequencies and not the frequencies themselves. For example, if the lowest frequency that can be used in a frequency band is *f1* and the highest is *f2*, then the bandwidth available is *f2 − f1*. Whereas wireless applications operate at a precise frequency, that frequency can vary within a band or range based on the several types of transmission methods employed by wireless LANs. Thus, the range of allowable frequencies represents the bandwidth supported by a particular application.

Power Measurements

Both the *bel* and *decibel* provide a mechanism to denote the difference between power transmitted and power received. Although we could express a difference between two signals it is often not as notable as the ratio provided by the bel and decibel. For example, suppose you transmitted a signal at 10 volts and received a signal at 1 volt. While we could say that the signal has a loss of 9 volts, it is probably more instructive to note that the original signal is only one-tenth as big at the receiver. Thus, a ratio can be more instructive than an absolute number.

When the telephone was deployed, a need arose to define the relationship between the received power of a signal and its original power. Initially that relationship was the bel (b), named in honor of Alexander Graham Bell, the inventor of the telephone. Although the preferred term used for power measurements changed to the decibel, the use of the bel and decibel is applicable not only to the one-hundred-year-old telephone but to modern wireless LANs. Let's examine the use of both terms.

BEL. The bel uses logarithms to the base 10 to express the ratio of power transmitted to power received. The resulting gain or loss for a circuit is given by the following formula:

$$B = \log_{10} \frac{P_0}{P_I}$$

Where B is the power ratio in bels, P_0 is the power output or received, and P_I is the input or transmitted power. The reason the bel uses logarithms for its measurement of power is because humans hear logarithmically. That is, the human ear perceives sound or loudness on a logarithmic scale. For example, if we estimate a signal to have doubled in loudness, the transmission power actually increased by approximately a factor of ten. Another reason for the use of logarithms in power measurements is the fact that changes to a signal in the form of signal boosts through amplifiers or signal loss caused by resistance are additive. The ability to add and subtract when performing power measurements based on a log scale simplifies computations. For example, a 10B signal that encounters a 5B loss and is then passed through a 15B amplifier results in a signal strength of $10 - 5 + 15$, or 20B.

Although the use of the bel dates to the beginning of telephony and transmission via wire, its use as a measurement of power received to power transmitted is applicable to nonwired applications including wireless LANs. Similarly, its cousin the decibel is also applicable as a power measurement for wireless LANs.

For those of us a bit rusty concerning the use of logarithms we can note that the logarithm to the base 10 ($\log_{10}$) of a number is equivalent to how many times 10 is raised to a power equal to the number. For example, $\log_{10}100$ is 2; $\log 10_{10}1000$ is 3, and so on. Because output or received power is normally less than input or transmitted power, the denominator in the prior equation is normally larger than the numerator. To simplify computations we can note a second important property of logarithms. That is,

$$\log_{10}\frac{1}{X} = \log_{10}X$$

As an example of the use of the bel for computing the ratio of power received to power transmitted, let us assume received power was one-hundredth of transmitted power. Then,

$$B = \log_{10}\left(\dfrac{\frac{1}{100}}{1}\right) = \log_{10}\dfrac{1}{100}$$

Because $\log_{10}\dfrac{1}{X} = -\log_{10}X$ we obtain:

$$B = -\log_{10}100 = -2$$

Note that a negative value indicates a power loss whereas a positive value indicates a power gain. Although the bel was used for many years to categorize the quality of transmission on a circuit, industry required a more precise measurement. This resulted in the adoption of the decibel (dB) as a preferred power measurement.

DECIBEL. The decibel (dB) represents the standard method used to denote power gains and losses. The dB is a more precise measurement because it represents one-tenth of a bel. The power measurement in dB is computed as follows:

$$dB = 10\log_{10}\left(\dfrac{P_0}{P_I}\right)$$

Where *dB* is the power ratio in decibels, P_0 is the output or received power, and P_I is the input or transmitted power. Returning to our previous example, where the received power was measured to be one-hundredth of the transmitted power, the power ratio in dB becomes:

$$dB = \log_{10}\left(\dfrac{\frac{1}{100}}{1}\right) = \log_{10}\dfrac{1}{1}$$

Because $\log_{10}\dfrac{1}{X} = -\log_{10}X$ we obtain:

$$dB = -\log_{10}100 = -20$$

We use different versions of the formulas for the bel and decibel when we compute power ratios and voltage ratios. The preceding equations used to compute the power ratios for bel and decibel change when we compute a voltage ratio. The reason for this can be explained by first returning to days of yore and remembering *Ohm's Law*. That fundamental law stated the relationship between current, voltage, and resistance as follows:

$$V = IR$$

Where V is voltage, I is current, and R represents resistance in ohms. Because power (P) is the product of current and voltage

$$P = IV$$

and $V = IR$. Power P is V^2/R. We can then express the voltage ratio as:

$$dB = 10\log_{10}(V_2^2/R)/(V_1^2/R)$$
$$= 10\log_{10}(V_2^2/V_1^2)$$
$$= 10\log_{10}(V_2/V_1)^2$$
$$= 20\log_{10}(V_2/V_1)$$

Thus, the formula for the ratio of voltage of a signal in dB becomes:

$$dB = 20\log_{10}(V_1/V_2)$$

DECIBEL ABOVE ONE MILLIWATT. It should be noted that the terms bel (B) and decibel (dB) do not indicate power. Instead, they represent a ratio or comparison between two power values, such as input and output power. Because it is often desirable to express power levels with respect to a fixed reference, it is common to use a 1-milliwatt (mw) standard input for comparison purposes. The use of 1 mw of input power dates to voice telephony. As a small piece of knowledge that might be useful for betting as you sit at a bar, 1 mw of power equals the average amount of speaker power of a typical telephone call. Thus, if you were wondering why telephone testing involves the use of a 1-mw test tone, as a famous radio announcer would say, "now you know the rest of the story." In the wonderful world of communications testing the 1-mw signal occurs at a frequency of 800 Hz. To ensure that we do not forget that the resulting power measurement occurred with respect to a 1 mw input signal, we use the term dBm. Here dBm becomes:

$$dBm = 10\log_{10}\left(\frac{10 \text{ Output power}}{1 \text{ mw input power}}\right)$$

and serves to remind us that the output power measurement occurred with respect to a 1-mw test tone. Although in most literature dBm is referred to as *decibel milliwatt,* in actuality it means *decibel above one milliwatt.* Thus, 10 dBm represents a signal 10 dB above or bigger than 1 mw, whereas 20 dBm represents a signal 20 dB above 1 mw and so on. Because a 30 dBm signal is 30 dB or 1,000 times larger than a 1 mW signal, 30 dBm is the same as 1 watt. We can use this relationship to construct the watts-to-dBm conversion table in Table 2.3.

TABLE 2.3

Relationship of
Watts and dBm

Power in Watts	Power in dBm
0.1 mW	−10 dBm
1 mW	−0 dBm
1W	−30 dBm
1 KW	−60 dBm

Signal-to-Noise Ratio

One of the more important metrics in communications is the *signal-to-noise (S/N) ratio*. In all communications systems there is a degree of noise, caused by the movement of electrons, powerline induction, and cross-modulation from adjacent wire pairs or, in wireless communications, from frequencies in adjacent channels. There are two basic categories of noise—thermal and impulse. *Thermal noise*, such as the movement of electrons or basic radiation from the sun, is characterized by a near uniform distribution of energy over the frequency spectrum.

Figure 2.3 illustrates thermal noise, which is also referred to as *white* or *Gaussian noise*. The selection of the name "Gaussian" results from the fact that the amplitude of thermal noise signals cumulatively follows a Gaussian distribution. This noise is also referred to as white noise because it contains all spectral frequencies equally on average, much as white light contains all of the colors of the rainbow.

Figure 2.3
White or thermal noise is characterized by a near uniform distribution of energy over the frequency spectrum.

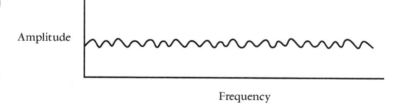

You can hear Gaussian or thermal noise if you tune your radio between stations. Doing so results in a hiss between stations. That hiss represents the Gaussian noise your receiver is pulling out of the air. One of the properties of noise worth mentioning is that it is proportional to bandwidth. This means that doubling the bandwidth of a channel does not quite double its capacity. In addition, because a radiated signal's power decreases inversely proportional to distance, Gaussian noise becomes more important as the distance between transmitter and receiver increases in a non-wired environment.

Because thermal noise is characterized by a near uniform distribution of energy over the frequency spectrum, it represents the lower level of sensitivity of a receiver. This is because a received signal must exceed the level of background noise if it is to be discriminated from the noise by the receiver. You can view the S/N ratios denoting the amount of unwanted electromagnetic noise relative to a signal's strength. As the level of noise power approaches the level of signal power it becomes harder to discriminate the signal from the noise. This results in a reduction in the ability of a transmission system to operate and can mean either a lower level of throughput or the inability to transfer data.

A second type of noise is formed from periodic disturbances, ranging from solar flares (commonly referred to as sunspots) to the effect of lightning and the operation of machinery. This type of noise is *impulse noise*. Impulse noise is illustrated in Figure 2.4. Note that impulse noise consists of irregular spikes or pulses of relatively high amplitude and short duration.

Figure 2.4
Impulse noise occurs at random times at random frequencies.

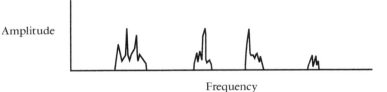

The signal-to-noise (S/N) ratio is used to categorize the quality of transmission. The S/N ratio is measured in dB and is defined as the ratio of the signal power (S) divided by the noise power (N) on a transmission medium. While you always want a S/N ratio greater than unity, because the receiver must be able to discriminate the signal from the noise, there are limits to how high the S/N ratio should be. Although a high S/N ratio is desirable, there are also limits associated with different systems that cap the allowable amount of signal power. These limits may be imposed by system operators or by the FCC or another regulatory body. Limits established by the latter are designed to minimize interference between different wireless systems.

In a wired transmission environment it is relatively easy to remove a good portion of potential noise. For example, through the use of a noise filter on an Ethernet LAN operating at 10 MHz it becomes possible to remove the potential occurrence of unwanted signals above the 10 MHz-signal rate. In a wireless environment it is more difficult to remove unwanted signals for several reasons. First, as an "open" system without shielding, wireless LANs are easily affected by other systems transmitting at or near the frequency of the wireless system. Second, the signal strength of most wireless systems including wireless LANs is relatively weak. In fact, the power of a wireless signal is inversely proportional to the square of the distance traveled. This means that a wireless signal dissipates at a significant rate as it radiates out in all directions. This also means that a low level of noise with respect to the transmitter in a wireless system can represent a much higher level of impairment if the source of the noise is closer to the receiver. A third and occasionally much more serious cause of noise, which we examine in more detail later, results from the reflections of signals and is referred to as *multipath distortion*. Such reflections result in one transmitted signal becoming multiple signals at a receiver.

Table 2.4 provides a summary of the relationship between dB and power or S/N ratios. The entries in Table 2.4 contain several values that are significant and deserve a degree of elaboration. First, consider a dB value of zero. Because the dB is defined as $10\log_{10}(P_o/P_I)$, this means that for a dB reading of zero $10\log_{10}(P_o/P_I)$ must be zero. This is only possible if P_o/P_I equals *unity*, which means that a dB value of zero occurs when the input power equals the output power. We can also note that a dB value of zero means there is no gain or loss at the termination point of a transmission system.

A second important dB value is 3, which is equivalent to a power or S/N ratio of 2:1. Thus, a 3 dB value indicates the signal is twice that of the noise. Finally, if you scan Table 2.4 and focus on dB values in multiple increments of 10, such as 100, 1000, and so on, you will note that they correspond to S/N ratios that increment by a power of 10. That is, a dB value of 10 is equivalent to a

TABLE 2.4

Relationship
between dB
and Power
Measurements

dB	Power or S/N Ratio
0	1.0:1
1	1.2:1
2	1.6:1
3	2.0:1
4	2.5:1
5	3.2:1
6	4.0:1
7	5.0:1
8	6.4:1
9	8.0:1
10	10.0:1
13	20.0:1
16	40.0:1
19	80.0:1
20	100.0:1
23	200.0:1
26	400.0:1
29	800.0:1
30	1000.0:1
33	2000.0:1
36	4000.0:1
39	8000.0:1
40	10000.0:1
50	100000.0:1

S/N ratio of 10; a dB value of 20 is equivalent to a S/N ratio of 100, and so on.

PROPAGATION LOSS. The loss in the strength of a signal as it traverses a medium is referred to as *propagation loss.* Propagation loss increases with respect to both distance and frequency. Concerning distance, as a signal propagates it encounters some type of resistance, which may include air molecules. Thus, the further a signal travels the weaker it becomes. Signal strength is inversely proportional to the square of the distance traveled. Thus, a doubling of transmission distance results in the reception of a quarter of a signal's original strength. Later in this chapter when we discuss transmission impairments, we will cover propagation loss in more detail, noting that in a wireless environment it is also referred to as *path loss*. The second factor that governs propagation loss is the frequency of a signal. If we remember our high school physics, we know that the high frequency components of a signal attenuate more rapidly than the low frequency components of a signal. Thus, propagation losses increase as the frequency used by a wireless system increases, all other things being equal. From an engineering and physics perspective, both propagation loss and frequency use explain why the area of coverage for older analog wireless systems is, in general, broader than the coverage provided by more modern PCS systems. Because PCS systems operate at approximately 1 GHz above analog systems (1.9 GHz versus 800 MHz), their signal loss is greater and the distance of coverage of a PCS cell is less than that of an analog wireless cell. In a wireless LAN environment the use of the 2.4 GHz ISM band results in a significant amount of propagation loss. When coupled with the fact that the power level of wireless LANs is limited to 1 watt in the United States and lower levels of power in some other countries, the ability of the signal to be received is significantly limited. This explains why the over-the-air transmission distance of a wireless LAN-compatible device is commonly limited to under 100 meters. This also explains why you commonly hear advertisements from radio stations stating that "WXYZ serves Outer Mongolia with 6 million watts of power!" To obtain the capability to have the broadcast received over a significant geographical area requires a high level of power.

Transmission Rate Constraints

There are two key constraints that govern the ability to transmit information at different data rates. These constraints are referred to as the *Nyquist Relationship* and *Shannon's Law*.

Nyquist Relationship

The Nyquist Relationship governs the signaling capability on a channel. In 1928, Harry Nyquist developed the relationship between bandwidth and the signaling rate (*baud*) on a channel as follows:

$$B = 2W$$

where *B* is the baud rate and *W* is the bandwidth in Hz.

BITS VERSUS BAUD. Prior to discussing what the Nyquist Relationship means in terms of the maximum achievable signaling rate on a channel, a brief digression into bits and baud is warranted. The *bit rate*, typically presented in terms of *bits per second* (bps), represents a measurement of data throughput. In comparison, baud represents the rate of signal change commonly expressed in terms of Hz. When information in the form of bits is to be transmitted, an oscillating wave is varied to impress or *modulate* information. The oscillating wave is referred to as a *carrier*. Common modulation techniques include altering the amplitude of the carrier (amplitude modulation), altering the frequency of the carrier (frequency modulation), and altering the time period or phase of the carrier (phase modulation). Some communications systems also alter two characteristics of the carrier, such as amplitude and phase.

We begin our examination of the relationship of bits and baud with a simple modulation scheme referred to as *frequency shift keying* (FSK). Here, each bit is modulated using one of two frequency tones, which we can refer to as f_1 and f_2. If we assume all binary 1s are modulated at f_1 and all binary 0s at f_2, or vice versa, this simple modulation scheme results in each bit's being equivalent to one signal change. Thus, in this situation the bit rate equals the baud rate.

Now let us assume a more sophisticated modulation technique. Using phase modulation (described in detail in Chapter 3) we can vary the phase of a signal according to the composition of a single bit or a group of bits. Suppose our phase modulation technique involves varying the phase of the carrier to one of four positions (0, 90, 180, and 270 degrees). This modulation technique then allows each possible combination of two bits to be encoded into one signal change. An example of this is shown in Table 2.5.

TABLE 2.5

Potential Mapping of Bit Pairs into Phase Changes

Bit Pair	Phase Change
00	0
01	90
10	180
11	270

In examining the relationship between bit pairs and phase change, it is clear that the bit rate is twice the signaling rate because two bits are packed into each signal change. The previously described technique is referred to as *dibit encoding* and represents one of many types of data modulation techniques. For now, the important aspect of this digression is to note that the bit rate may or may not equal the baud rate, with equality depending on the method of data encoding used by a modulation scheme. Some types of wireless systems, such as the Home RF system, also modulate voice. This is because the human spectrum of speech is primarily concentrated from approximately 300 Hz to 3300 Hz, an area of the frequency spectrum that requires rather long antennas. To use higher portions of the frequency spectrum, voice is modulated, using the well-known *amplitude modulation* (AM) broadcasts on radio. Now that we have an appreciation for the difference between bit and baud as well as the need to modulate both data and voice for wireless transmission, let us return to our discussion of the Nyquist Relationship.

MAXIMUM MODULATION RATE. The Nyquist Relationship states that the maximum rate at which data can be transmitted prior to one symbol's interfering with another, a condition referred to as *intersymbol interference*, must be less than or equal to twice the bandwidth in Hz. Because most transmission systems modulate or vary a signal, the Nyquist Relationship puts a cap on the signaling rate that is proportional to available bandwidth. In the wonderful world of modem design, this explains why designers use different techniques to pack more bits into each baud to achieve a higher data transfer rate: Because the bandwidth of a telephone channel is fixed, a modem designer cannot exceed a given signaling rate prior to intersymbol interference adversely affecting the data flow. Thus, to transmit more data the modem designer must pack more bits into each signal change. A second constraint, however, puts a cap on the maximum achievable data rate that can be obtained. That constraint is Shannon's Law.

SHANNON'S LAW. In 1948, Claude E. Shannon presented a paper concerning the relationship of coding to noise and computed the theoretical maximum bit rate capacity of a channel of bandwidth W Hz. The relationship developed by Shannon is:

$$C = W\log_2\left(1 + \frac{S}{N}\right)$$

Where:
    C = capacity in bits per second (bps)
    W = bandwidth in Hz
    S = power of the transmitter
    N = power of thermal noise

In 1948, a "perfect" telephone channel was considered to have a S/N ratio of 30 dB, which represents a value of 1000. The maximum data transmission for a telephone channel in 1948 was then proposed as:

$$C = W\log_2\left(1 + \frac{S}{N}\right)$$

Where:
    $W = 3000 \log_2(1 + 103)$

$$S = 3000 \log_2 (1001)$$
$$N \cong 30000 \text{ bps}$$

It should be noted that over 50 years after Shannon's paper, the maximum data rate on an analog telephone channel is 33.6 Kbps, which is within 10 percent of Shannon's Law. In actuality, telephone channels today have a bit—no pun intended—less noise. Although you are probably well aware of 56 Kbps modems, they only operate near that rate downstream when the destination location has a direct digital connection. In the upstream direction the 56 Kbps modem is still limited to a data transmission rate of 33.6 Kbps.

Now that we have an appreciation for the roles of the Nyquist Relationship and Shannon's Law in governing signaling and data rates, let us turn our attention to the frequency spectrum because this is the medium used by wireless LAN systems.

Radio Frequency Spectrum Allocation

On a worldwide basis there are authorities in most countries that operate as government agencies and regulate the use of the frequency spectrum. Under the provisions of the International Telecommunications Union (ITU) treaties with most countries, these countries are obligated to comply with the radio frequency spectrum allocations specified by the ITU for international use. This ensures that aircraft can contact ground stations, satellites can transmit TV signals, and cellular subscribers can use their phones without encountering interference from other signals as country borders are approached. ITU treaties also permit each country to allocate the domestic use of the frequency spectrum differently from the international allocation, as long as the domestic allocation does not conflict with neighboring country frequency spectrum allocation. This means that you can expect slight to major differences in the domestic use of the frequency spectrum between countries, especially when you travel from one continent to another where differences become more pronounced. In this section we primarily focus our attention on the allocation of radio frequency in the United States. As we briefly describe and discuss different types of wireless LAN systems in this chapter and in more detail in subsequent chapters, however, we will reference, when applicable, frequency use in different areas of the world.

U.S. Spectrum Allocation

In the United States the Communications Act of 1934, as revised, resulted in the authority for managing the use of the radio frequency spectrum partitioning between the U.S. Commerce Department's National Telecommunications and Information Administration (NTIA) and the Federal Communications Commission (FCC). The NTIA administers the frequency spectrum for Federal government use. In comparison, the FCC, which is an independent regulatory agency, administers the frequency spectrum for non-federal government utilization. The managed radio frequency spectrum currently ranges from 9 KHz to 300 GHz. This spectrum is divided into over 450 frequency bands.

In examining the allocation of the frequency spectrum in the United States, we will not look at every frequency band. Instead, we will primarily focus our attention on bands applicable to wireless LANs and on other bands that can be used to provide a general frame of reference for several popular applications, such as AM and FM radio, different television bands, and certain types of satellite communications systems. First, however, let us examine the relationship between frequency band nomenclature and the frequencies to which they refer. Doing so will provide us with a reference for noting the band nomenclature associated with different types of wireless applications.

BAND NOMENCLATURE.   Over the past century several methods were devised to categorize the frequency spectrum. One of the more popular methods uses the wavelength as a power of 10 metric for denoting each particular frequency band. Figure 2.2 represents an example of categorizing different frequency bands based on wavelength. If you look carefully at the right portion of Figure 2.2 you will note that the *ultra low frequency* (ULF) band is more easily categorized as signals with a wavelength from 10^8 to 10^7 meters rather than by frequency. Similarly, the *extremely low frequency* (ELF) band represents the frequency spectrum where wavelength varies from 10^7 to 10^5 meters, whereas the *very low frequency* (VLF) band represents the frequency spectrum where the wavelength varies from 10^6 to 10^4 meters, and so on. Table 2.6 provides a summary of well-known frequency bands and their wavelength range, in meters, based on powers of 10.

TABLE 2.6

Well-Known Frequency Bands

Wavelength Frequency band	(Meters)
Ultra Low Frequency (ULF)	10^8—10^7
Extremely Low Frequency (ELF)	10^7—10^5
Very Low Frequency (VLF)	10^5—10^4
Low Frequency (LF)	10^4—10^3
Medium Frequency (MF)	10^3—10^2
High Frequency (HF)	10^2—10^1
Very High Frequency (VHF)	10^1—1
Ultra High Frequency (UHF)	1—10^1
Super High Frequency (SHF)	10^1—10^2
Extremely High Frequency (EHF)	10^2—10^3
Electro Optical Frequency (EOF)	10^3—10^8
High Energy Frequency (HEF)	10^8—10^{13}

Another popular method used to categorize the frequency spectrum comes from one of the well-known regulators of the frequency spectrum, the FCC. The FCC categorizes the frequency spectrum from 0 to 400 GHz by noting the overlapping of bands that fall into a range of frequencies. Table 2.7 provides a summary of the FCC's frequency band nomenclature, indicating the frequency range associated with distinct bands as well as certain pairs of bands. If you compare the entries in Table 2.7 to the entries in Figure 2.2, you will note that although there is a correspondence between the two, it is not an exact correspondence. Many times the frequency range of a band or of a particular application is referenced, rather than the band nomenclature, to ensure both parties are on the same wavelength—no pun intended. Now that we have a general appreciation for the frequency spectrum and general knowledge of different frequency bands, let's look at existing and emerging wireless LAN applications and their use of the frequency spectrum.

TABLE 2.7

FCC Frequency
Band
Nomenclature

Frequency Band	Frequency Range
Very Low Frequency/Low Frequency (VLF/LF)	0—130 KHz
Low Frequency/Medium Frequency (LF/MF)	130—505 KHz
Medium Frequency (MF)	505—2107 KHz
Medium Frequency/High Frequency (MF/HF)	2107—3230 KHz
High Frequency (HF)	3230—28000 KHz
High Frequency/Very High Frequency (HF/VHF)	33—162.0125 MHz
Very High Frequency/Ultra High Frequency (VHF/UHF)	162.0125—322 MHz
Ultra High Frequency (UHF)	322—2655 MHz
Ultra High Frequency/Super High Frequency (UHF/SHF)	2655—3700 MHz
Super High Frequency (SHF)	3700 MHz—27.5 GHz
Super High Frequency/Extremely High Frequency (SHF/EHF)	27.5—32 GHz
Extremely High Frequency (EHF)	32—400 GHz

Applications

To obtain a general appreciation of the range of wireless applications and the diverse band of frequencies allocated to representative applications, I skimmed through the one-inch-thick summary of FCC frequency allocations in the United States and extracted twenty. Those twenty applications (including their general frequency block) are listed in Table 2.8.

TABLE 2.8

Common and Evolving Wireless Applications

Application	Frequency
AM Radio	535—1635 KHz
Analog Cordless Telephone	44—49 MHz
Television	54—88 MHz
FM Radio	88—108 MHz
Television	174—216 MHz
Television	470—806 MHz
Wireless Data (to be licensed)	700 MHz
RF Wireless Modem	800 MHz
Cellular	806—890 MHz
Digital Cordless	900 MHz
Personal Communications	900—929 MHz
Nationwide Paging	929—932 MHz
Satellite Telephone Uplink	1610—1626.5 MHz
Personal Communications	1850—1990 MHz
Industrial, Scientific and Medical	2400—2483.5 MHz
Satellite Telephone Downlinks	2483.5—2500 MHz
Large Dish Satellite TV	4—6 GHz
National Information Infrastructure	5.15—5.25 GHz
Small Dish Satelite TV	11.7—12.7 GHz
Wireless Cable TV	28—29 GHz

If you examine the entries in Table 2.8 you will note there are two wireless LAN frequency blocks reserved by the FCC. Both frequency bands represent unlicensed spectrum; however, there are power regulations associated with transmission in each frequency block. The ISM block from 2.400 through 2.4835 GHz supports a transmission of 1W in the United States. This frequency block is used by the IEEE 802.11 and 802.11b standards that operate at 1, 2, 5.5, and 11 Mbps; these are covered in considerable detail later in this book. The 2.4 GHz band is used by a diverse series of applications, some of which can be expected to cause interference when operated close to a wireless LAN operating in that band. The second frequency band shown in Table 2.8 is split into three subbanks and provides a total of 300 MHz of spectrum. The first 100MHz, from 5.15 GHz to 5.25 GHz is restricted to a maximum power output of 50 mw. The second 100 MHz permits a more generous 250 mw of power. The third 100 MHz is delegated for outdoor applications and permits a maximum power output of 1 W. The unlicensed *National Information Infrastructure* (NII) or *UNII* band will be used by products complying with the IEEE 802.11a specifications, which permits data rates up to 54 Mbps.

Other Transmission Impairments

We conclude this chapter by examining several additional transmission problems that are associated with the use of wireless LANs. In doing so we primarily focus on indoor wireless communications because a majority of wireless LAN applications occur inside buildings. The transmission impairments we will describe, however, are also applicable to outdoors. In comparison, the opposite may not be true. For example, some outdoor wireless applications can encounter such transmission impairments as sun spots and thunderstorms, and may even be unlucky enough to experience the effect of Johnny Backhoe digging where he is not supposed to be. These impairments primarily effect outdoor wireless and wired WANs much more than they do an indoor wireless LAN.

Basic Wireless LAN Components

Let's start our discussion of wireless LAN impairments with a schematic illustrating a few components of a wireless LAN. Figure 2.5, which illustrates the use of a wireless LAN in a small building, represents the starting point of our discussion.

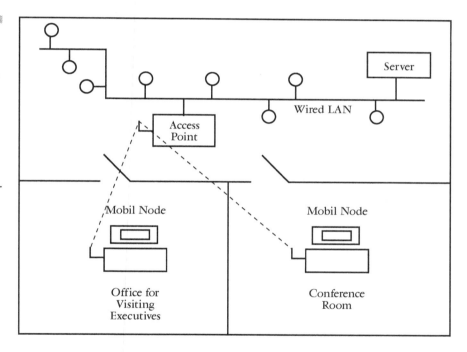

Figure 2.5
In an office environment a wireless LAN typically connects mobile nodes in conference rooms and other temporary locations via one or more access points to a wired LAN.

In examining Figure 2.5, note two wireless mobile nodes are shown installed in a conference room and an office area. In comparison, wall partitions are used to provide office working areas to other employees whose terminals are hardwired to an existing LAN. The access point represents a transmitter-receiver, which is referred to as a *transceiver*. The access point's primary function is to connect wireless nodes to the wired network at a fixed location. The access point receives data from the wired LAN, buffers it, and transmits data to the mobile nodes. In the opposite direction, the mobile nodes transmit to the access point, which temporarily stores data in its buffer and, when it gains access to the wired LAN, transmits information onto the fixed network. A single access point can support a small group of mobile nodes, with the number of nodes supported and the range between the nodes and the access point based on several factors including the network operating rate, position of antennas, and obstructions in the line of sight between the nodes and the access point. Although the wireless LAN nodes are labeled "mobile nodes," in most situations they are only mobile until the laptop or notebook computer is positioned at an appropriate location. At that point the mobile node commonly remains relatively fixed with respect to its location.

Now that we have an appreciation for the manner by which a wireless LAN can be used in an office environment, let us turn our attention to several common transmission impairments we must consider.

Path Loss

The first factor that affects the ability of a signal to reach and be recognized by a receiver is path loss. Path loss, *attenuation,* and *propagation loss* are synonymous. Path loss represents the signal attenuation between transmit and receive antennas and is primarily a function of the transmission distance between transmitter and receiver. Other factors, including obstructions in the line of sight between transmitter and receiver and the wavelength of the signal, can also affect the strength of a signal at a receiver.

In a wireless transmission environment path loss (P_L) or signal attenuation between transmit and receive antennas is computed as follows:

$$P_L = P_t - P_r + G_t + G_r$$

Where:

P_t represents transmit power in dB
P_r represents received power in dB
G_t represents transmit antenna gain in dB
G_r represents receiver antenna gain in dB

In *ideal free space*, where there are no reflections and a signal radiates or spreads out in all directions evenly, the path loss is proportional to the square of the separation distance (d) between transmit and receive antennas. According to research performed at a government laboratory in Boulder, Colorado, we can compute path loss in free space (P_{fs}) as follows:

$$P_{fs} = 10n\log_{10}\frac{(4d)}{\wedge}$$

where $\wedge$ is the wavelength, d is the distance of separation between transmit and receive antennas, and a value of 4 for n instead of 2 provides a more realistic consideration of a signal-cluttered environment. Because each indoor environment can be expected to differ from another with respect to signal clutter, we employ a simple model to illustrate the effect of signal attenuation or path loss. This model shows the loss as proportional to the square of the separation distance (d^2). Thus, at a distance of 10 meters, the received signal strength represents 1/100 or one-hundredth of the original signal strength:

$$dB = 10\log_{10}\frac{P_0}{P_I}$$

Thus, at a distance of 10 meters the power ratio becomes:

$$dB = 10\log_{10}\frac{\frac{1}{100}}{1} = 10\log_{10}\frac{1}{100}$$

because $\log_n \dfrac{1}{X} = -\log_n X$ we obtain:

$$dB = -10\log_{10}100 = -20$$

Now let us consider what happens when the distance between the transmitter and receiver increases to 100 meters. At a distance of 100 meters the received signal strength is now $1/100^2$ or one-ten-thousandth of the original signal. Thus, at a distance of 100 meters, and again assuming no reflections, the power ratio becomes:

$$Db = 10\log_{10} \dfrac{\dfrac{1}{10000}}{1} = 10\log_{10} \dfrac{1}{10000}$$

because $\log_n \dfrac{1}{X} = -\log_n X$ we obtain:

$$dB = -10\log_{10}10000 = -40$$

Although it is beyond the range of low powered wireless LANs, let us also investigate the power ratio at a distance of 1000 meters. At that distance the radiated signal strength at the receiver would be $1/1000^2$ or one-millionth of the original signal. Thus, the power ratio becomes:

$$dB = 10\log_{10} \dfrac{\dfrac{1}{1000000}}{1} = -10\log_{10}1000000 = -60$$

We can plot the path loss in terms of the previously computed power ratios, again considering neither the effect of reflections nor other signals within the geographic area. Figure 2.6 illustrates a general dB plot of path loss by distance.

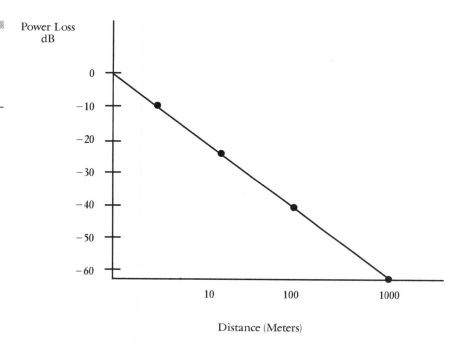

Figure 2.6
Path loss by distance between transmit and receive antennas.

In examining Figure 2.6, it is important to remember that the dB represents a logarithmic value. Thus, −60 dB at 1000 meters means that one-millionth of the signal power is received. This also means that because the transmission power permitted for wireless LANs is capped at 1W in the United States, the sensitivity level of a receiver must be pretty darn good as the distance between transmitter and receiver increases. Unfortunately, it becomes very expensive to design receivers with a sensitivity level beyond −40 to −45 dB. In addition, even though receivers can be designed to operate at those levels it becomes difficult to discriminate a signal from noise. Because of this, most wireless LANs support a transmission distance of up to approximately 100 meters.

Now that we have an appreciation for the effect of path loss on transmission distance, let us turn our attention to another related topic that affects the received signal—multipath propagation.

Multipath Propagation

In an indoor environment a radiated signal can be expected to collide with many types of surfaces. In addition to partitions and walls, signals encounter desks, chairs, ceilings, and even people. Both human and nonhuman objects can reflect portions of a signal flowing from a transmitter to a receiver with a resulting condition referred to as *multipath propagation.*

Figure 2.7 illustrates the potential effect of multipath propagation on a signal radiating from a transmitter towards a receiver. In this example two surfaces are shown reflecting portions of the radiated wave. For simplicity, only a portion of the radiated transmitted wave is shown.

Figure 2.7
Multipath propagation results from the reflection of signals off different types of surfaces.

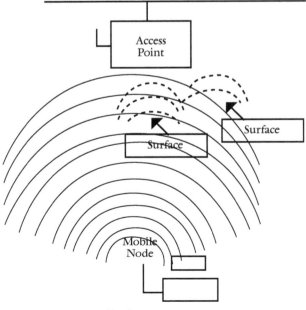

Legend - - - - Reflections

In examining Figure 2.7, note that the signal radiating from the mobile node is shown being reflected from two surfaces. Note that those radiated, reflected signals arrive at the antenna of the access point slightly delayed from the portion of the original signal. In addition, if you are familiar with the manner by which radar operates, you are probably aware that when the friendly highway patrol officer points the radar gun at your vehicle the time and frequency shift of the radar reflections are used to denote the speed of your vehicle. That is, as radar bounces off your vehicle a frequency shift occurs. Similarly, as the mobile node's transmission is reflected off different surfaces a frequency shift of the reflected signal occurs. Thus, multipath propagation creates both a time and frequency dispersion at certain locations that can adversely affect the quality of the signal being received. Although Figure 2.7 illustrates the effect of multipath propagation when transmission occurs from a mobile node to an access point, the reverse is also true: When an access point transmits to a mobile node, any intermediate surfaces can cause signal reflections that also cause time and frequency signal problems.

One of the major problems of multipath propagation is the fact that a receiver appears to have multiple signals. Because reflections appear as an overlap of modulated signals this results in intersymbol interference and increases the *bit error rate* (BER) at the receiver. The actual degree of intersymbol interference caused by multipath propagation depends on the number of reflective surfaces in the general area where the mobile nodes and access point are located, the angle the signals are reflected from, the surfaces they encounter, and the distance from the surfaces and the transmitter to the receiver. Because a signal is weaker as the distance increases between transmitter and receiver, the effects of multipath propagation is normally more pronounced at the receiver.

Fading

Fading represents an obstruction to a signal. If the signal obstruction is in a direct line between a transmitter and receiver it is possible that the obstruction blocks the signal from being received.

Because RF signals radiate outward, however, it is more likely that weak reflections arrive at the receiver, resulting in a weak or faded signal.

Enhancing Signal Reception

Multipath propagation results in signals shifted in both time and frequency whereas fading results in a weaker received signal. Because we discuss modulation methods in detail in Chapter 3, we only briefly describe here techniques we can use for enhancing signal reception. As you might expect, however, the best method to enhance signal reception—boosting signal strength— is not an option because the power of a wireless LAN signal is regulated by FCC rules. Because manufacturers do not want to violate federal regulations, the next best technique is either to alter the position of the antenna or to change the operating frequency. While it is relatively easy to reposition an antenna, once again FCC rules prohibit the arbitrary changing of frequency. Frequency changing is still an option (and in fact occurs) if the changes are within a prescribed band or range of frequencies allocated by the FCC to wireless LAN operations. Frequency hopping, which is described in Chapter 3, can alleviate a portion of interference from multipath propagation. This is because a frequency hopping receiver uses the same pseudorandom code to change frequencies in tandem with the transmitter. Because the original signal is received prior to reflected signals, arriving at a receiver, frequency hopping can reduce a portion of the adverse effect of multipath propagation. Because frequency hopping occurs based on a pseudorandom number code, however, it is also possible for the next selected frequency to approach a shifted reflected signal's frequency. Because of this, frequency hopping can, on rare occasions, result in a receiver's positioning itself to a frequency that matches a reflected signal.

Another technique that can be used to minimize the effect of multipath propagation is to transmit multiple carriers, each at a lower signal power than the other. This action, which is referred to as *multitone, orthogonal frequency division multiplexing* (OFDM), or *coded OFDM* (COFDM) (discussed in Chapter 4), minimizes the data signaling rate per carrier. This in turn minimizes the signal delay when obstructions cause reflections.

Perhaps the easiest method to consider using in an attempt to overcome the effect of multipath propagation involves antenna positioning. In addition to positioning an antenna to receive the highest level of signal power, another technique that can be used involves multiple antennas. Referred to as *antenna diversity*, this technique involves the separation of antennas by an odd multiple of a quarter wavelength, which at 2.4 GHz is approximately 7.5 inches or 3 cm. There are several techniques used with antenna diversity. One technique is for a receiver simply to select the highest level of signal strength, switching from one antenna to the other as necessary. Another technique is to combine signals received from multiple antennas to generate an optimum signal-to-noise ratio. While antenna diversity is commonly used for fixed wireless transmission, its cost generally precludes its employment in low cost LAN adapters. Thus, the technology of wireless LANs that involves the use of frequency hopping and other transmission techniques provides a built-in mechanism against a good portion of multipath propagation impairment. Some common sense is always useful in setting up an access point and mobile nodes: Positioning the antennas correctly and moving unnecessary obstacles along the path between nodes and an access point can go a long way towards enhancing your ability to communicate.

Understanding Wireless LAN Modulation

In this chapter, we will obtain an understanding of how binary data generated by laptops, notebooks, servers, and other digital devices are transported through the air in a wireless LAN environment. We commence with the basics, assuming no prior knowledge of modulation schemes. We first examine the three main types of modulation and see how each modulation method is used to impress information on what is referred to as the carrier signal. Once this is accomplished we turn our attention to how several modulation methods applicable to wireless LANs operate, and we conclude with an overview of wireless LAN transmission methods, using these to include different modulation techniques.

Basic Modulation Methods

In this section we examine how data is modulated for transmission. Prior to doing so, however, let us refresh our memory concerning the rationale for modulation and what the term means.

Rationale

If you consult a dictionary you will note the term *modulate* is defined as "to vary, regulate or alter" typically followed by the term "signal." While that definition is technically correct in describing the modulation process, it does not tell us why we modulate information. So, let us start at the beginning and discuss the basic reason for this process.

Modulation represents a technique used to alter the characteristics of a signal. As we alter the signal we do so in a manner that conveys information to a receiver. Thus, the primary purpose of the modulation process is to convey information.

Modulation is applicable to both analog and digital transmission systems. Many readers probably use modems to access the Internet by making a call over the *public switched telephone network* (PSTN). In doing so the modem modulates digital signals generated by your computer into analog signals suitable for transmission over the PSTN, which is an analog transmission facility. The reverse process, referred to as *demodulation*, converts an analog signal back into its digital form and is also performed by modems on received data. If we use Integrated Service Digital Network (ISDN) or even an Ethernet or Token-Ring wired LAN, each time we transmit information, the unipolar digital signals generated by our personal computer are converted or modulated into a different type of digital signals for transmission. The reason for this digital-to-digital modulation results from the fact that certain types of signals are more suitable than others for conveying information and for obtaining a reasonable transmission distance. Regardless of whether modulation is digital-to-analog or digital-to-digital, its goal is to alter the characteristics of a signal to convey information. This said, let us turn our attention to the modulation process.

Modulation Process

As previously noted, the purpose of modulation is to alter the characteristics of a signal to impress information on the signal. In the field of radio frequency (RF) communications the signal is referred to as a carrier. By itself a carrier represents a repeating signal that conveys no information—well, almost no information. For example, if you turn on your modem's speaker and hear the high pitched noise of a carrier signal prior to its being modulated, you at least know there is a path between two devices. Thus, prior to impressing information, the presence of a carrier signal informs you that you have continuity.

To impress information on a carrier requires one or more of the characteristics of the carrier signal to be altered. For analog signals the carrier is commonly a *sine wave*, represented by the following formula:

$$a = A\sin(2\pi ft + \varnothing)$$

where *a* represents the instantaneous value of voltage at time *t, A* represents the maximum amplitude, *f* represents the frequency of the signal, and ∅ represents the phase of the signal. Thus, the characteristics of a carrier signal that can be altered include its amplitude for *amplitude modulation* (AM), its frequency for *frequency modulation* (FM), and its phase for *phase modulation* (∅M). As we will note later in this chapter, we can also alter two or more signal characteristics during the modulation process. But before we jump from the frying pan into the fire, let us start our examination of modulation by focusing our attention on the three basic methods used to alter a carrier.

Amplitude Modulation

Amplitude modulation represents a very simple modulation method that can be used to impress information onto a carrier signal. Under the amplitude modulation process the magnitude of the signal is varied from a zero level, used to represent a binary 0, to a fixed peak-to-peak voltage level to represent a binary 1.

Figure 3.1 illustrates an example of amplitude modulation. The top portion of Figure 3.1 illustrates digital data encoded in a unipolar nonreturn to zero (NRZ) format. Under this coding method, which is used by terminal devices including PDAs, PCs, and servers, a binary zero is represented by a lack of voltage whereas a binary 1 is represented by a positive voltage. Note that succeeding binary 1s result in the digital signal's staying at a high voltage—thus the "nonreturn to zero" portion of the name of the digital signaling method. The lower portion of Figure 3.1 illustrates the use of amplitude modulation to encode the digital data stream shown in the upper portion of the figure into an appropriate sequence of analog signals. While amplitude modulation is relatively easy to perform, it is normally used for very low data rates. Otherwise, at high data rates, when used by itself, a transmission impairment is easily mistaken for modulation that does not exist. When amplitude modulation is used in conjunction with phase modulation, however, it becomes more difficult for noise to be misinterpreted for both a phase change and an amplitude change. Because of this, amplitude modulation is commonly used with phase modulation to obtain a high-speed data modulation method, a topic we investigate later in this chapter.

Figure 3.1
The modulation of data using amplitude modulation results in a zero level representing a binary 0 and a fixed peak-to-peak voltage representing a binary 1.

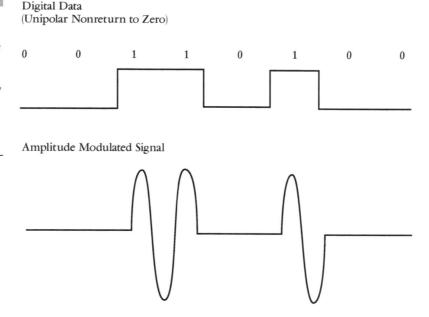

Digital Data
(Unipolar Nonreturn to Zero)

0   0   1   1   0   1   0   0

Amplitude Modulated Signal

Frequency Modulation

A second method that can be used by itself to modulate data is frequency modulation. As we noted in Chapter 2, frequency represents the rate at which a signal repeats itself. Thus, by changing the frequency of a signal to correspond to the binary 1s and binary 0s of a digital data source we can impress information onto a carrier.

One of the earliest uses of frequency modulation can be traced to low-speed modems manufactured during the 1970s. Such modems employed a transmitter that shifted a carrier signal between two tones or frequencies based on whether the data to be modulated was a binary 1 or a binary 0.

Figure 3.2 illustrates frequency modulation. In this example the frequency is shown varied or shifted between one frequency (F_1) and a second frequency (F_2) which represents frequency shift modulation.

Figure 3.2
Frequency modulation results in the variance of the frequency of a carrier signal in tandem with changes in the composition of digital data.

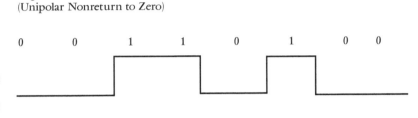

Digital Data
(Unipolar Nonreturn to Zero)

0   0   1   1   0   1   0   0

Frequency Modulated Carrier Signal

f_1 f_2 f_1 f_2 f_1

The term *frequency shift modulation*, while correct, is more commonly referred to as *frequency shift keying* (FSK) modulation. The term *keying* dates to the manner by which legacy telegraph systems operated. A Morse key operator would hold down his key for a short period of time to represent a "dot," while the key would be held down for a longer period to denote a "dash." For movie experts, the "V" for victory signal in the John Wayne D-Day epic was created via three dots followed by a dash. In any event, when a Morse key was depressed by the telegraph operator, it resulted in an electrical contact that turned the transmitter on and transmitted a carrier frequency wave. When the key was raised the transmitter was turned off and the carrier wave terminated. Thus, the carrier shifted between an on and off condition in tandem with spaces between dots and dashes.

During the early part of the twentieth century the introduction of automatic telegraph systems resulted in the radio-frequency carrier's being on all the time, with information transmitted by the shifting of the carrier upward in frequency to represent a mark or binary 1, while the frequency of the carrier was shifted downward in frequency to represent a space or binary 0. Thus, the automation of the telegraph resulted in the addition of "shift" keying, because the carrier was shifted. Later, with the development of modems, FSK became a very popular modulation method for data rates up to 1200 bps. By convention, probably dating to the automation of the telegraph, a binary 0 is assigned to the lower frequency while a binary 1 is assigned to the higher frequency.

Phase Modulation

A third method employed to modulate data is phase modulation. Phase modulation represents the process of varying a carrier signal with respect to the origination of its cycle.

THE SINE WAVE. The better to understand phase modulation let us use the sine wave many of us remember from physics. That wave starts at 0 degrees, peaks at 90 degrees, returns to 0 at 180 degrees, turns negative and reaches a maximum negative value at 270 degrees, and returns to 0 at 360 degrees. The top portion of Figure 3.3 illustrates one cycle of a standard sine wave.

VARYING A CARRIER. We can vary the phase of a carrier by delaying the wave cycle. That is, if we delay the wave by a quarter of its cycle the carrier will be 90 degrees out of phase. Similarly, if we delay the signal by a half a cycle, it will be 180 degrees out of phase. The lower portion of Figure 3.3 illustrates a sine wave shifted 180 degrees out of phase.

Under phase
modulation the
phase or position of
the wave is altered to
impress information
onto a signal.

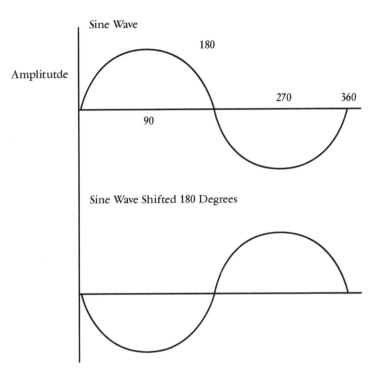

Sine Wave

Amplitutde

180

90

270

360

Sine Wave Shifted 180 Degrees

SINGLE-BIT PHASE MODULATION. There are several methods of phase modulation. In its simplest form, referred to as *single-bit phase modulation,* a transmitter simply shifts the phase of a signal between two phases to represent a binary 1 and a binary 0. This type of signaling is referred to as *phase shift modulation.* Because the transmitter "keys" itself to one of two phase shifts this modulation technique is also referred to as phase shift keying (PSK).

When modems were designed for operation on the public switched telephone network engineers had a limited bandwidth of approximately 4000 Hz they could use. If you remember our discussion of the Nyquist Relationship presented in Chapter 2, the signaling rate is limited to twice the bandwidth of a channel prior to one signal's adversely affecting another, a situation referred to as intersymbol interference. Thus, with 4000 Hz of bandwidth a signal could oscillate at a maximum rate of 8000 baud before intersymbol interference occurring. If modem designers only packed one bit into each signal change, the maximum data rate obtainable over the public switched telephone network would be 8 Kbps, which we know is not true. Thus, the question arises: How do modern modems that operate on the PSTN achieve data transmission rates beyond 8 Kbps? The answer to this question is that PSTN modem designers pack more than one bit into each signal change. For example, consider the use of a transmitter that shifts a signal into one of four phase angles. If the modem designer packs two bits into each possible signal change the data rate becomes twice the baud or signaling rate. This type of data coding is referred to as *dibit encoding*. Similarly, assume a modem designer develops a technique that allows a transmitter to shift the phase of a signal into one of eight positions. This allows three bits to be encoded into one signal change and results in a data transmission rate three times that of the signaling rate. As you might expect, this modulation method is referred to as *tribit encoding*.

MULTILEVEL PHASE-SHIFT KEYING.   You should also note that because two or more bits are encoded into a phase shift, both dibit and tribit encoding into phase shifts is referred to as *multilevel phase-shift keying.* "Multilevel" denotes the fact that dibit coding has four possible phase shifts, each representing two bits, while tribit encoding has eight possible phase shifts, each representing three bits. "Phase shift" indicates that the phase shifts jump from one to another or can be considered as keyed to a certain angle. Table 3.1 provides three examples of possible phase angle shifts for dibit and tribit coding techniques.

It would appear that we could continue packing more bits into each phase angle to obtain higher data rates. That is, if three bits per baud is good, certainly four is better, five still better, and six begins to excel. Unfortunately, this is not true for reasons that will soon become apparent.

Consider the left portion of Figure 3.4, which illustrates the subdivision of a sine wave represented by a circle into four phase changes. This type of diagram is referred to as a spatial representation of phase modulation. Now consider the right portion of Figure 3.4, which illustrates the subdivision of the sine wave into eight phase changes. Note that the phase angle values decrease by a factor of two as the number of possible phase angle shifts doubles. As phase changes decrease, it becomes harder for a receiver to discriminate one phase shift from another. In addition, any transmission impairment that rotates a phase shift just a small amount has a better chance of causing a transmission error as the number of phase shifts increases. Because of this it is common practice to combine phase and amplitude to produce modulation signal points at discreet locations as well as to limit the number of phase shifts.

TABLE 3.1

Common Phase-Angle Values Used in Multilevel, Phase-Shift Keying Modulation

Coding Technique	Bits Transmitted	Possible Phase (degrees)		Angle Shifts
Dibit	00	0	45	90
	01	90	135	0
	10	180	225	270
Tribit	000	0	22.5	45
	001	45	67.5	0
	010	90	112.5	90
	011	135	157.5	135
	100	180	202.5	180
	101	225	247.5	225
	110	270	292.5	270
	111	315	337.5	315

Figure 3.4
Spatial representation of four- and eight-level phase shift keying modulation.

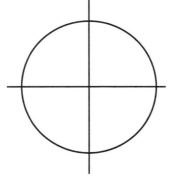
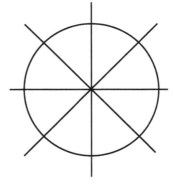

M-ARY OPERATION. The presentation of phase changes in Figure 3.4 represents a common method used to indicate phase changes. When all possible signal points are displayed, the resulting diagram is referred to as a *signal constellation point diagram*. Another term that warrants attention is *M-ary*, which refers to the density of the constellation pattern as well as the number of phase shifts in a modulation scheme. To illustrate M-ary operations let us begin with a single carrier tone for which we assign binary 0 to a 0° phase change and a binary 1 to a 180° change. This type of system is referred to as a *binary phase shift key* (BPSK) modulation method whose spatial relationship is shown in Figure 3.5.

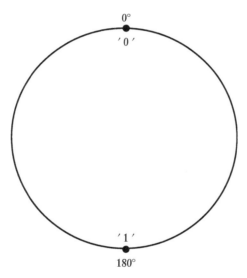

Figure 3.5
A spatial representation of BPSK modulation.

If we encode pairs of bits into one of four phase changes we achieve the ability to pack two bits into each signal (phase) change. The four phase changes result in a M-ary modulation scheme with M = 4 and represent a technique referred to as *quadrature phase shift keying* (QPSK). Figure 3.6 illustrates a spatial representation of QPSK modulation, where the phase shifts correspond to the first column to the right of dibit coding in Table 3.1.

Figure 3.6
A spatial
representation of
QPSK modulation
where M = 4 for a
M-ary operation.

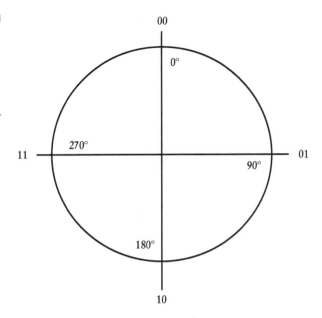

If we return to Figure 3.4, we note that the left portion of the illustration is equivalent to a M-ary value for M of 4, whereas the right portion of the figure has a value of M = 8. As M increases, the angle difference between phase shifts decreases, making it more difficult for a receiver to discriminate one signal from another.

COMBINED MODULATION.   One of the most popular modulation techniques combines amplitude and phase modulation. One example of a combined modulation method involves the encoding of each sequence of four bits into an amplitude and phase change. This technique is referred to as *quadrature amplitude modulation* (QAM). One rather dated example of QAM, which nevertheless provides a background concerning how the process operates, is the consultative committee for International Telephone and Telegraph (now ITU–T) V.29 modem standard. The V.29 modulation method results in the first bit in each sequence of four bits defining the relative amplitude of the modulated signal whereas the following three bits define its phase. Table 3.2 lists the relative signal element amplitudes of a V.29 QAM process based on the value of the first bit in each quadbit. The value of the following three bits defines the phase change. For example, the quadbit of 1100 would have a signal amplitude of 5 and a phase change of 270 degrees. The resulting plot of all possible signal values is shown in Figure 3.7.

TABLE 3.2

V.29 Signal Amplitude Determination

Absolute Phase	First Bit	Relative Signal Element Amplitude
0, 90, 180, 270	0	3
	1	5
45, 135, 225, 315	0	$\sqrt{2}$
	1	$3\sqrt{2}$

Figure 3.7
The V.29 signal constellation pattern.

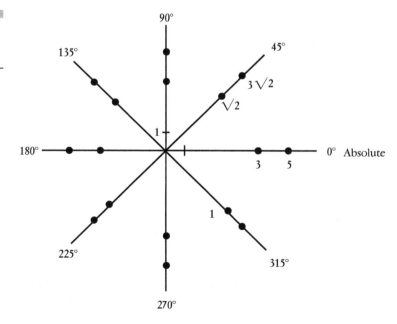

If you take each dot shown in Figure 3.7, the collection of all possible signal points is referred to as the *signal constellation* pattern. A good modem designer structures the points as far away as possible from one another to limit the effect of transmission impairments that may result in the receiver's misinterpreting one combination of phase shift and amplitude for another. This misinterpretation occurs if a signal point moves more than 50 percent from the angle of the phase shift, because the receiver automatically assigns bits to the received signal during the demodulation process based on matching the received signal point to the closest point in the signal constellation. Thus, a good modem designer seeks to develop a constellation pattern where points in the pattern are as far away from one another as possible. Unfortunately, as more bits are encoded into each signal change the density of the points increases. Thus, a point is reached where the distance between points becomes too close and the resulting modem might only be suitable for a laboratory environment.

Now that we have an appreciation for QAM, let us turn our attention to radio frequency (RF) modulation methods used in a wireless LAN environment.

Wireless LAN Modulation Methods

In the first section in this chapter we focused our attention on modulation basics, examining how amplitude, frequency, and phase can be used by themselves or paired to modulate a digital data stream. In this section we turn our attention to radio frequency (RF) modulation techniques used by wireless LANs.

In examining wireless modulation methods we group them based on the wireless spread spectrum techniques employed. The first spread spectrum technique that we will examine is the *direct sequence spread spectrum* (DSSS) broadband transmission method. We briefly examined DSSS in Chapter 1, and it will be described in considerable detail later in this book. But for now, let's focus our attention on DSSS modulation.

DSSS Modulation

There are three modulation methods supported under DSSS. The original IEEE 802.11 standard specified the use of *differential binary phase shift keying* (DBPSK) at a 1 Mbps data rate. Using DBPSK, each bit is mapped into one of two phases, resulting in a 1 MHz baud rate. To provide a 2 Mbps data rate, DSSS requires the use of *differential quadrature phase shift keying* (DQPSK), which operates on pairs of bits referred to as *dibits*. Because this technique packs two bits into each signal change, it permits a signaling rate of 1 MHz to support a data rate of 2 Mbps.

The first extension to the IEEE 802.11 standard, the 802.11b specification, provides support for data rates of 5.5 Mbps and 11 Mbps. Both data rates use quadrature phase shift keying (QPSK) modulation, which we examine in this section.

Differential Binary Phase Shift Keying (DBPSK)

Using differential binary phase shift keying (DBPSK), each data bit is mapped into one of two phases, with information encoded based on the phase difference between adjacent data symbols.

DBPSK uses a phase change of 0 degrees to represent a 0 bit and a phase change of 180 degrees to represent a 1 bit. For an illustration of the operation of DBPSK modulation, consider Figure 3.8. The first portion of the figure (the extreme left) represents the reference phase. If we assume the digital sequence of 0011 is transmitted, the first two modulations do not alter the reference phase because a 0-degree phase change is used to represent a binary 0. The third modulation results in a phase change of 180 degrees added to the reference phase, resulting in a 180-degree phase change. Then, the fourth modulation, which also encodes a binary 1, adds 180 degrees to the preceding phase, resulting in a 360-degree phase change that restores the original signal because 360 degrees is equivalent to 0 degrees. Thus, we note that the differential phase portion of the term DBPSK can be denoted as:

$$\varnothing_n = \Delta\varnothing + \varnothing_{n-1}$$

where:
   $\varnothing_n$ = transmitted phase for bit n
   $\varnothing_{n-1}$ = transmitted phase for bit n-1
   $\Delta\varnothing$ = phase change

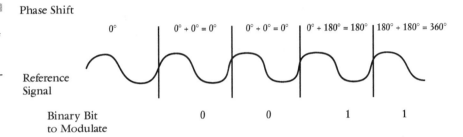

Figure 3.8
DBPSK modulation of the bit sequence 0011.

Differential Quadrature Phase Shift Keying (DQPSK)

A second modulation method employed by wireless LANs is differential quadrature phase shift keying (DQPSK). DQPSK is also commonly used with cellular systems and by cable modems.

DQPSK is similar to DBPSK in that the transmitted phase of the coded data is a function of the previous phase and the phase change as noted earlier in this chapter. However, the "Q" in DQPSK denotes that this modulation technique encodes data by mapping pairs of bits into one of four distinct phase changes, resulting in M = 4 for a M-ary operation. Table 3.3 lists the mapping of dibits into phase changes under DQPSK modulation. Note that this mapping is the same as that shown in the first column in Table 3.1 for dibit coding.

TABLE 3.3

DQPSK Bit
Encoding

Dibit	Phase Change
00	0
01	90
10	180
11	270

Figure 3.9 illustrates an example of DQPSK modulation. In this example the data sequence 00, 01, 10, and 11 is assumed to have occurred so that each pair of bits resulted in a phase change based on the phase changes noted in Table 3.3.

Figure 3.9
DQPSK modulation
example.

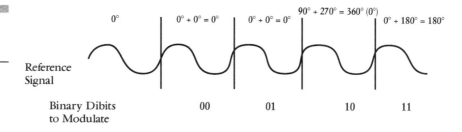

In examining Figure 3.9, note that we begin with the in-phase reference signal shown to the extreme left portion of the diagram. The first dibit (00) results in a 0-degree phase shift, thus the signal is not shifted from the reference signal. Next, the dibit of 01 results in a 90-degree phase shift, which when added to the previous shift of 0 degrees results in a 90-degree shift of phase. The third dibit (10) adds 270 degrees of phase shift to the previously 90-degree-shifted signal. This results in a total shift of 360 degrees, which restores the signal to its reference unshifted phase shown at the extreme left of the illustration. The fourth dibit (11) requires a 180-degree differential phase change to the prior signal. This results in the signal's being shifted 180 degrees out of phase from the prior signal. Because the prior signal was 360 degrees out of phase, which returned it to its original reference signal position, the signal to modulate the dibit of 11 is out of phase by 0 degrees + 180 degrees or a total of 180 degrees.

Complementary Code Keying (CCK) QPSK

A third modulation method supported by DSSS is only applicable to the IEEE 802.116 standard that provides data rates of 5.5 Mbps and 11 Mbps. Complementary code keying (CCK) represents a form of vector modulation that uses complex symbol structures as a mechanism to spread a signal. Thus, although referred to as a modulation method, CCK more accurately represents a spreading function. CCK consists of a set of 64 8-bit code words, with a structure that provides a unique mathematical property that enables different code words to be distinguished from one another even under extreme levels of noise and multipath interference. To achieve a 5.5 Mbps transmission rate, CCK encodes 4 bits per signal change, using quadrature phase shift keying (QPSK) modulation operating at a 1.375 MHz rate. To obtain an 11 Mbps transmission rate, CCK encodes 8 bits per signal change and also uses QPSK as the modulation method at a baud rate of 1.375 MHz. Table 3.4 summarizes the four data rates, modulation techniques, signaling rates, and number of bits encoded per baud for the 802.11b wireless LAN. In examining the entries in Table 3.4, note that the 1 Mbps and 2 Mbps rates are supported by the original 802.11 specification, whereas the 5.5 Mbps and 11 Mbps operating rates represent the extensions supported by the 802.11b specification.

TABLE 3.4

IEEE 802.11b Modulation

Data Rate	Modulation	Baud Rate	Bits/Baud
1 Mbps	BPSK	1 MHz	1
2 Mbps	QPSK	1 MHz	2
5.5 Mbps	QPSK	1.375 MHz	4
11.0 Mbps	QPSK	1.375 MHz	8

The IEEE 802.11b specification is limited to supporting DSSS. Thus, to examine the full range of modulation methods used by different IEEE 802.11 standards we must focus on two different communications methods supported by the family of 802.11 standards: *Frequency hopping spread spectrum* (FHSS) and *orthogonal frequency division multiplexing* (OFDM).

Frequency Hopping Spread Spectrum (FHSS) Modulation

FHSS represents a second type of RF communications supported by the original IEEE 802.11 standard issued during 1977. FHSS results in the split of the available frequency spectrum into a series of small subchannels. Using a pseudorandom number algorithm, a transmitter hops from subchannel to subchannel, transmitting short bursts of data on a subchannel prior to jumping to the next subchannel. (In Chapter 4, when we cover wireless LAN communications systems, we will cover the operation of FHSS in detail.)

FHSS operates at either 1 or 2 Mbps, with its data rate based on the use of *Gaussian frequency shift keying* (GFSK) modulation.

Gaussian Frequency Shift Keying (GFSK)

GFSK represents a variation of frequency shift keying (FSK). FSK represents one of the earliest methods used to modulate data. Using FSK, a carrier tone is shifted or keyed upward in frequency to represent a mark or binary 1 and downward in frequency to represent a space or binary 0. The term *shifting* refers to the fact that the carrier is shifted to one of two frequencies to modulate each bit. By convention, binary 1 is assigned the higher frequency and binary 0 is assigned the lower frequency.

A GFSK modulator is the same as an FSK modulator, shifting between frequencies based on the data to be modulated. Prior to bits' entering the modulator, however, they are passed through a

Gaussian filter. For better understanding, another digression and a review of Gaussian noise and Gaussian filtering is in order.

Gaussian noise represents random noise and is commonly called white noise because it contains all spectral frequencies equally on average, similar to the manner in which white light contains all of the colors of the rainbow. While the level of Gaussian noise is typically uniformly low at a transmitter and receiver, the power of a signal decreases inversely proportional to the square of the distance between devices. This means that the effect of Gaussian noise is more pronounced at the receiver. Thus, filtering such noise prior to modulation results in a better received signal. To accomplish this a GFSK modulator passes broadband pulses through a *Gaussian filter*. The objective of the filter is to make each pulse smoother, limiting its spectral width. This technique is also referred to as *pulse shaping*, because it alters the shape of each pulse.

To illustrate the manner in which Gaussian filtering occurs, assume a pulse jumps from 0 to 1. This action forces the modulated waveform to change rapidly, resulting in a large out-of-band spectrum. Instead of allowing the pulse to change rapidly, the Gaussian filter smoothes the pulse, letting it rise in increments to modulate the carrier. This has the effect of reducing the out-of-band spectrum.

Returning to our discussion of FHSS modulation methods, to obtain a 1 Mbps data rate, two-state GFSK is used. To obtain a 2 Mbps data rate, four-state GFSK is employed. Now that we have an appreciation for the modulation methods supported by FHSS let us turn our attention to the third method of wireless LAN communications. That method is known as *orthogonal frequency division multiplexing* (OFDM).

Orthogonal Frequency Division Multiplexing (OFDM) Modulation

OFDM is the communications method defined for use in the second extension to the IEEE 802.11 standard known as the 802.11a specification. This specification supports data rates up to 54

Mbps. The key to the ability to achieve a higher data rate is the fact that each subchannel in the 5 GHz band used by the 802.11 specification supports multiple carriers, where the carriers are orthogonal and independent to each other.

Figure 3.10 illustrates the relationship between an 802.11a channel and its eight subchannels. Each subchannel consists of fifty-two narrow-band carriers, each 312.5 KHz in width. Forty-eight of these carriers are used to transport data and four are used to provide pilot tones. A total of 16.875 MHz of bandwidth is used in each 20 MHz channel, with the difference employed as a guard band between channels.

Figure 3.10

Relationship between an 802.11 channel and its eight subchannels.

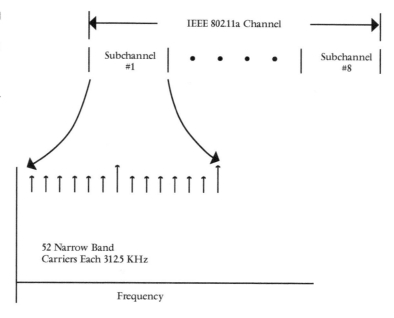

Each of the subcarriers in a subchannel is modulated using binary phase shift keying (BPSK), quadrature phase shift keying (QPSK), or one of two versions of quadrature amplitude modulation (QAM). Because we previously discussed BPSK and QPSK, in the remaining portion of this section we will primarily describe and discuss the use of QAM using OFDM. Prior to doing so, however, it is worth noting that the goal of this chapter is to describe RF modulation methods and not describe the communications systems used by RF modulation. Although we briefly described DSSS, FHSS, and OFDM in this chapter, we go into considerable detail in Chapter 4 when we describe and discuss communications systems used to transport data. That said, let us turn our attention to QAM.

Quadrature Amplitude Modulation (QAM)

The need for speed required the IEEE to take advantage of the greater bandwidth of the 5 GHz frequency spectrum allowed for use by the 802.11a extension to the 802.11 standard. In addition to supporting BPSK and QPSK the 802.11a standard added 16-QAM and 64-QAM.

The use of 16-QAM results in four bits coded into a distinct signal point. Using 16-QAM, there are four signal points per quadrant or sixteen points for the full constellation pattern.

Figure 3.11 illustrates the signal constellation pattern for 16-QAM specified for use by the 802.11a standard. Note that the constellation pattern is Gray-coded. The *Gray code*, which is shown in Table 3.5 for a 3-bit code, has the property that, between any two successive binary numbers only one bit changes state.

TABLE 3.5

Binary and Gray Code Comparison

Decimal	Binary	Gray Code
0	000	000
1	001	001
2	010	011
3	011	010
4	100	110
5	101	111
6	110	101
7	111	100

Figure 3.11
Constellation pattern
for 16-QAM 802.11a
modulation.

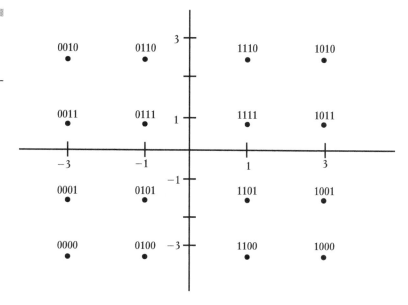

Figure 3.11
Constellation pattern for 16-QAM 802.11a modulation.

In 16-QAM operation, because data is fed into a convolutional coder that adds redundancy, the coding rate and the number of code bits per subcarrier determine the number of coded bits per OFDM symbol and the number of data bits per OFDM symbol. Similarly, 64-QAM results in six bits coded into each signal change and uses sixteen points per quadrant or a total of sixty-four signal points in its constellation pattern. The resulting data rate is governed by the coding rate of the convolutional coder as well as the number of code bits per carrier and the number of subcarriers.

16-QAM. The IEEE 802.11a standard specifies 16-QAM using a half- or three-quarter-coding rate, each with four coded bits per subcarrier. Because there are forty-eight subcarriers used to transport data, this results in 48 x 4 or 192 coded bits per OFDM symbol.

When a half coding rate is used, the number of data bits per OFDM symbol becomes 192 × .5 or 96. In comparison, when a three-quarter-coding rate is used, the number of data bits per symbol becomes 192 × $3/4$ or 144.

64-QAM. The key difference between 16-QAM and 64-QAM is in the number of coded bits per subcarrier. Using 16-QAM, four bits are coded per subcarrier. In comparison, using 64-QAM, six bits are coded per carrier.

A second difference between 16-QAM and 64-QAM concerns the coding rates supported. As we noted in our discussion of 16-QAM, that modulation method is used with half- and three-quarter-coding rates. In 64-QAM, coding rates of three-quarters and two-thirds are used. Table 3.6 provides a summary of the modulation methods, coding rates, coded bits per OFDM symbol, data bits per OFDM symbol, and nominal bit rate for the IEEE 802.11a standard.

TABLE 3.6

IEEE 802.11a
Modulation
Methods and
Operating
Parameters

Modulation	Coding Rate	Coded bits/ Subcarrier	Coded bits/ OFDM symbol	Data bits/ OFDM symbol	Nominal Data Rate (Mbps)
BPSK	1/2	1	48	24	6
BPSK	3/4	1	48	36	9
QPSK	1/2	2	96	48	12
QPSK	3/4	2	96	72	18
16-QAM	1/2	4	192	96	24
16-QAM	3/4	4	192	144	36
64-QAM	3/4	6	288	216	54
64-QAM	2/3	6	288	192	48

In examining the entries in Table 3.6, note that the 6, 12, and 24 Mbps data rates are mandatory for all IEEE 802.11a products. Thus, 9, 12, 18, 36, 48, and 54 Mbps represent optional data rates that may or may not be supported. Because 802.11a products had not yet reached the market when this book was written, it may be a while before I can comment on vendor support for the range of data rates defined by the standard.

Wireless LAN Communications Systems

Although a modulation process is required to transmit data in a radio frequency environment, the manner in which the process occurs is directed by the communications system used. In this chapter we examine the operation of three communications systems that are used by current and emerging wireless LANs. We will examine frequency hopping spread spectrum (FHSS), direct sequence spread spectrum (DSSS), and a special version of orthogonal frequency division multiplexing (OFDM) referred to as *coded orthogonal frequency division multiplexing* (COFDM). Each represents a wireless LAN communications system that governs the manner in which information is transmitted.

Spread Spectrum Communications

Although the focus of this chapter is on the operation of three distinct communications systems used in wireless LANs, two are directly related to spread spectrum communications, so let us first turn our attention to obtaining an appreciation of the rationale for the development of this technology. In doing so we will also note one of those interesting facts of life and discuss how a movie star many associate with beauty and charm was the intellect behind the development of one method of spread spectrum communications.

Development Rationale

Three spread spectrum communications methods were developed during the 1940s as a mechanism to provide a reliable and secure communications method for the military during battlefield conditions. In a battlefield environment enemy forces use radio frequency (RF) receivers to scan the frequency spectrum for the transmissions of their opponent. Once these were located, the enemy could either attempt to listen to the transmissions if they were in the clear, attempt to break any coded transmission, or jam a transmission. In a rapidly evolving military operation, communications were normally encoded, which left little time to attempt to decode intercepted messages. Therefore, jamming was the primary method used to disrupt the communications infrastructure. Because communications occurred at a fixed frequency the enemy only had to locate the frequency being used and transmit noise or gibberish at a higher power level than the power level used by the opponent. Even if the opponent switched to a back-up frequency it was relatively easy to locate the secondary frequency and jam transmission again.

General Operation

In spread spectrum communications more bandwidth is used than in conventional narrowband transmission that is based on the use of a specific radio frequency. This results in a signal that is louder and easier to detect. Because the signal is spread over a greater frequency spectrum, however, the receiver must know the parameters associated with the spread-spectrum signal being broadcast to recover the transmitted signal correctly. Otherwise, if the receiver is not tuned to the correct frequency, the spread-spectrum signal appears as static or background noise. Thus, the use of spread-spectrum transmission makes communications more difficult to jam because the enemy needs to know the parameters of the spread-spectrum signal or attempt to jam a wide range of frequencies, the latter a daunting task. In addition, if the parameters of the spread-spectrum signal are not known, then it is essentially impossible to intercept more than a small portion of communications as an enemy scans the frequency spectrum.

Spread-Spectrum Methods

There are three general methods that can be used to obtain a spread-spectrum communications capability. Those methods include frequency hopping spread spectrum (FHSS), direct sequence spread spectrum (DSSS), and time-hopped spread spectrum (THSS). Because the first two techniques are used in wireless LANs, we limit our coverage of spread spectrum in this chapter to those two communications methods.

FREQUENCY HOPPING SPREAD SPECTRUM (FHSS). In frequency hopping spread spectrum a narrowband signal moves or hops from one frequency to another using a pseudorandom sequence to control hopping. This results in a signal's lingering at a predefined frequency for a short period of time, which limits the possibility of interference from another signal source generating radiated power at a specific hop frequency.

Figure 4.1 illustrates a plot of frequency hopping spread spectrum communications in a time and frequency domain. Note that the desired signal hops from one frequency to another at a set power, which is normally regulated by the Federal Communications Commission (FCC) in the United States.

Figure 4.1
Frequency hopping spread spectrum results in a signal hopping from one frequency to another.

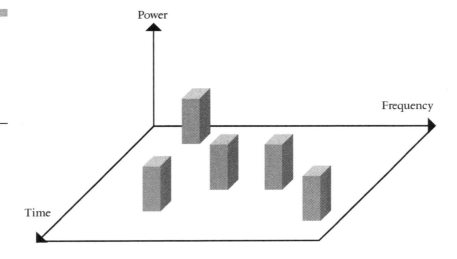

ORIGINATION.   If you are a movie fan and remember watching the classic flick *Samson and Delilah* you probably remember the actress Hedy Lamar. Although she never received an Oscar for her acting, this Vienna-born beauty can be considered as the father, or better yet, the mother of FHSS communications.

As a trophy wife of a German industrialist she visited armament factories and viewed field tests of torpedo systems. When she fled Germany she looked for ways to assist her newly adopted country when war came to North America. Working with the composer George Antheil, she was discussing the ease with which a radio signal sent to control a torpedo could be blocked; in a moment of inspiration she realized that hopping from one frequency to another provided a mechanism to overcome jamming. Their patent for a "Secret Communications System" was granted on August 11, 1942. Although never implemented because of the specification for the use of piano rolls to provide frequency hopping synchronization, the invention of the transistor allowed the concept of a beautiful actress to bear fruit. As a famous radio announcer would say, "Now you know the rest of the story."

UTILIZATION.   FHSS represents a popular wireless communications method. In addition to its use in wireless LANs, FHSS is employed by Bluetooth, a short-range communications system that may have hundreds of millions of participants within a few years. In the military, FHSS is commonly used to overcome potential jamming, and the technology is also used in a modified form in satellite and cellular communications.

Direct Sequence Spread Spectrum (DSSS)

A second type of spread spectrum communications used by wireless LANs is direct sequence spread spectrum (DSSS). Under DSSS a random string, referred to as a spreading code, is used to map a small number of data bits into a larger number of bits that are modulated and spread across a wide frequency band.

GENERAL OPERATION. Figure 4.2 illustrates the general operation of DSSS. In examining Figure 4.2, note that because transmission is spread across a wide frequency band as a result of the spreading process, transmission power of the spread signal is reduced. Although the lower power makes the signal susceptible to noise, the spreading code adds redundant data that provides a mechanism to recover the original signal. Of course, data bits are spread via the use of a spreading code, which, when modulated, results in a greater use of frequency at a particular point in time. Thus, you can collocate more FHSS networks within an area than when using DSSS networks. While this may tend to favor the use of FHSS when multiple wireless LANs are required within close proximity to one another, it is also important because a receiver must know the same spreading sequence to be able to correctly receive the spread transmission.

Figure 4.2
Direct sequence spread spectrum uses a spreading code that spreads a signal over a wide frequency spectrum.

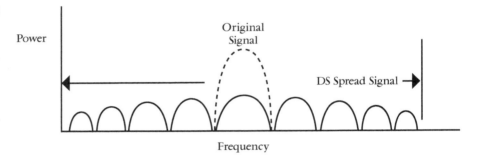

COMPARISON TO FHSS. Although both FHSS and DSSS represent spread spectrum technologies, they use two different approaches to communicate. In FHSS, a signal hops from one frequency to another, remaining at each frequency for a short period of time. In DSSS, another group of FHSS networks must be placed at a relatively long distance from the first group to prevent hopping interference. In comparison, if we are grouping DSSS networks, two such groups can be located closer to one another because of the redundant coding used to minimize the effect of interference. Thus, there is no simple answer to which represents a better technology. Fortunately, because most organizations require one or only a few collocated wireless LANs, a direct comparison between the two may not be necessary. However, because some organizations will require the use of a significant number of collocated wireless LANs, as we examine FHSS and DSSS we can obtain the detailed information necessary to compare and contrast their use.

Frequency Hopping Spread Spectrum

In frequency hopping spread spectrum (FHSS) communications a narrowband carrier is shifted in discrete increments of frequency. The frequency shift is based on a pattern generated by a pseudo-random algorithm that spreads the transmission over a wide frequency band: thus the name of this communications technique.

Regulations

Earlier we noted that the use of the radio frequency spectrum is regulated. In the United States the regulating body is the FCC. FCC rules define applications that can use certain frequency bands, the communications system available for use in the band, and the transmit power that can be used within a designated band. In addition, depending on the type of communications system permitted, the FCC may have additional rules that define the manner in which the communications system operates.

Until September 2000 frequency hopping was only permitted in the unlicensed 900 MHz band within the 902 to 928 MHz range. A dozen vendors developed equipment for operation within that band, but such products were proprietary. In September 2000 the FCC amended its rules to permit FHSS in the 2.4 GHz band. The purpose of this rule change was to allow wider bandwidths that would enable wireless communications at higher data rates. Under Section 15.247 of FCC rules a new subparagraph (a)(1)(iii) was added that stated:

> "Frequency hopping systems in the 2400–2483.5 MHz band may utilize hopping channels whose 20 dB bandwidth is greater than 1 MHz provided the systems use at least 15 nonoverlapping channels. The total span of hopping channels shall be at least 75 MHz. The average time of occupancy on any frequency shall not be greater than 0.4 seconds within a 30-second period."

In addition to adding the above subparagraph, section 15.247's subparagraph (b)(1) was revised to read as follows:

> "For frequency hopping systems in the 2400–2483.5 MHz employing at least 75 hopping channels, all frequency hopping systems in the 5725–5850 MHz band, and all direct sequence systems the maximum power is 1 watt. For all other frequency hopping systems in the 2400–2483.5 MHz band: 0.125 watts."

Based on these rulings the FCC specified a number of parameters by which a FHSS system must operate at in the 2400 GHz band. For example, a FHSS system must use at least 75 hopping channels, with each channel spaced 1 MHz from the other. In addition, because the average time of occupancy on any frequency (technically known as the *dwell time*) cannot exceed 0.4 seconds in a 30-second period, the time before a frequency is allowed to repeat is 30 seconds. In addition, note that FCC rules also regulate the transmit power for FHSS. Thus, the number of channels, duration at a particular channel, and when a frequency can be repeated, as well as the allowable power, is regulated.

Table 4.1 compares the general characteristics of FHSS in the 900 MHz and 2.4 GHz bands. Note that the larger bandwidth in the 2.4 GHz band permits a higher data rate to be achieved. In addition, the standardization of the IEEE 802.11 specification enables vendors to produce compatible products in the 2.4 GHz band and this fosters interoperability.

TABLE 4.1

FHSS Band
Comparison

Frequency	Time before Repeating a Frequency	Number of Channels
902—928 MHz	20 seconds	50
2.4—2.4835 GHz	30 seconds	75

Operational Parameters

Because the FCC regulates the many FHSS operational parameters that govern the manner by which frequency hopping occurs, we can revise Figure 4.1 by eliminating power from consideration because its value is fixed by the FCC. Redrawing frequency hopping spread spectrum in the frequency and time domain results in the diagram shown in Figure 4.3. Because it is important to ensure that all of the channels in a hopping pattern are used before channels in the pattern are reused, a minimum hopping rate must be determined. With a dwell time of 400 ms and a minimum of 75 channels this means that the time to hop through all channels becomes 400 ms/channel × 75 channels or 30 seconds. Thus, the minimum hopping rate is 75 hops/30 seconds or 2.5 hops/second.

Figure 4.3
Frequency hopping spread spectrum communications vary the carrier frequency in discrete increments based on a predefined code sequence or algorithm.

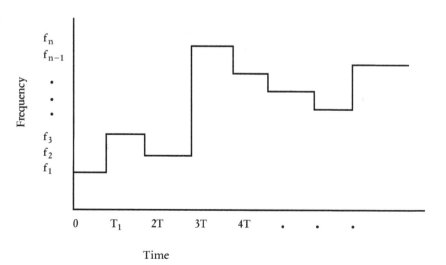

Packet Transmission Capability

With a dwell time of 400 ms it is possible to transmit a large number of maximum length Ethernet packets at one frequency prior to hopping to the next frequency. For example, consider the maximum length Ethernet frame, which is 1526 bytes in length including an 8-byte prefix used for synchronization. This is equivalent to 1526 bytes × 8 bits/byte or 12208 bits. At a data rate of 1Mbps the transmission time required to transfer a maximum length Ethernet frame becomes 12208 bits/1000000 bps or 12.208 ms. Thus, in a 400 ms dwell time it becomes possible to transmit 400 ms/12.208 ms or approximately 32 maximum length Ethernet frames.

A second operating rate supported by FHSS is 2 Mbps. At a 2 Mbps operating rate the transmission time required to place a maximum-length Ethernet frame into the air becomes 12208 bits/frame/2 Mbps or 6.104 ms. Again using a dwell time of 400 ms, it becomes possible to transmit 400 ms/6.104 ms or 65 maximum length Ethernet frames.

Hopping Modes

In examining Figure 4.3 note that the signal frequency remains constant for a specified time duration, which is referred to as the dwell time. We can compute the hopping rate of a FHSS communications system. The rates between the hopping rate expressed in hops per second (hps) and the data rate expressed frames per second (fps) can be used as a mechanism to define two distinct FHSS modes. When the hopping rate occurs at a rate that is faster than the frame rate, the FHSS communications system is referred to as a *fast hop* or *fast frequency hopping system*. This means there is not enough dwell time to transmit a frame, thus requiring two or more hops for a frame to be transmitted. In comparison, when frequency hopping occurs at a hop rate slower than the frame rate, the FHSS communications system is referred to as a *slow hop* or *slow frequency hopping system*. When this occurs it becomes possible to transmit one or more frames at one frequency prior to hopping to the next frequency.

In a military environment *forward error correction* (FEC) can be used with a fast hop spread-spectrum technique to minimize the effect of interference from natural causes or jamming as well as to correct errors caused by natural causes or jamming. In comparison, a slow frequency hopping system enables at least one frame to be transmitted at a hop frequency and uses retransmission as a mechanism to correct any bit errors occurring in a frame. In a wireless LAN environment slow frequency hopping is employed.

Advantages of Use

There are a number of advantages associated with the use of frequency hopping spread spectrum communications. One of its key advantages is its ability to overcome interference. To understand how frequency hopping minimizes the potential effect of interference, consider Figure 4.4, which illustrates an example of a frequency hopping sequence amid some background noise that could result from the use of a cordless phone, microwave oven, a Bluetooth-compliant PDA, or another device operating in the 2.4 GHz band. In examining Figure 4.4, note that the hop sequence shown is from 1 to 3, 3 to 7, 7 to 5, 5 to 2, 2 to 4, 4 to 6, and 6 to 9. Because noise is shown to peak around channel 2 in the frequency spectrum, data is transmitted on channel 1, and when a hop occurs to channel 3, more data is transmitted. Only when the frequency hop occurs from channel 5 to channel 2 is noise at a sufficient level for an adverse affect on the transmission of data. Because the dwell time is 0.4 seconds, however, quicker than you can say Jack Flash, a hop to channel 4 occurs and the new transmission channel is at a frequency outside of the area of noise.

From a technical perspective, the term *processing gain* (PG) is used to measure the performance advantage of spread spectrum communications versus narrowband communications. As we noted earlier, FHSS obtains its wideband modulation characteristics by switching its narrowband signal over a wide range of frequencies in time. In a frequency hopping environment the processing gain is defined as the ratio of the bandwidth of each hop to the bandwidth of the transmission channel in dB. If we return to the example shown in Figure 4.4, it should be obvious that the more instantaneous the change in frequencies and the wider the overall bandwidth for changing frequencies, the higher the immunity to interference. Thus, a lower PG represents more immunity to interference than a higher processor gain has.

Figure 4.4
Frequency hopping spread spectrum minimizes the effect of noise.

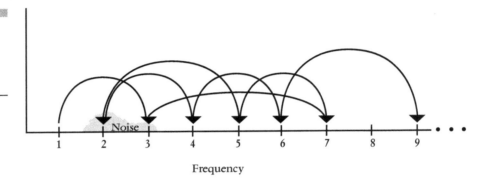

Another advantage of FHSS is the fact that it is possible to deploy multiple systems within a relatively close geographic area, because there are 79 usable channels in the 2.4 GHz band. This means that FHSS channel usage between two systems has a 1 in 79 probability (or approximately 1 percent) of overlap. Figure 4.5 illustrates how two systems can coexist and transmit using different hop sequences that minimize interference between systems.

A third advantage associated with FHSS concerns its immunity to multipath interference. As a brief refresher, multipath interference results from signals bouncing off walls, desks, doors, and even persons, resulting in multiples of a signal arriving at the receiver at different times. FHSS design provides a solution to multipath interference because a FHSS system hops to a different frequency—literally a step ahead of the reflections.

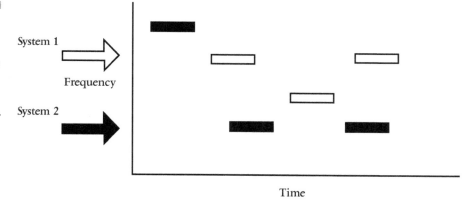

Figure 4.5
Nonoverlapping
channels minimize
interference between
FHSS systems located
in close proximity to
one another.

Direct Sequence Spread Spectrum

Direct sequence spread spectrum (DSSS) represents a second method of spread spectrum communications used by wireless LANs. In DSSS, data entering a transmitter is combined with a higher data rate bit sequence, commonly known as a *chipping code.* The chipping code is known to both the transmitter and receiver and forms the basis for spreading data bits over a wide range of frequencies. Because the chipping code represents a pseudorandom sequence that results in the spreading of data over a wide band that resembles a higher level of noise, this code is also referred to as a *pseudonoise* (PN) code. Each bit in the code is called a *chip* to note that they are used to encode and decode data bits.

Regulation

Much as with FHSS, the FCC regulates many aspects of DSSS communications. Those aspects of DSSS regulated by the FCC include the frequency band where DSSS can be used, the maximum transmit power allowed, and the bandwidth permitted.

DSSS operations for wireless LANs can occur in the Industrial, Scientific, and Medical (ISM) bands described in Chapter 2 we noted that these bands can vary. In the United States ISM bands consist of the 902 to 928 MHz, 2.4000 to 2.4835 GHz, and 5.725 to 5.850 GHz frequencies. Because the 2.4 GHz frequency band is available in most countries, it was selected for the original IEEE 802.11 wireless LAN, which specifies both frequency hopping spread spectrum and direct sequence spread spectrum as options available for use at either 1 or 2 Mbps. (Later in this book we examine the IEEE 802.11 standard and its two extensions, referred to as 802.11a and 802.11b.)

Operation

The chipping code can be viewed as a pseudorandom signal operating at a higher rate than the actual data rate. Because the transmission rate is proportional to bandwidth, the net effect of the use of a chipping code is to spread the resulting signal over a greater frequency spectrum.

The actual combination of data bits and the bits in a chipping code occurs through the use of modulo-2 addition. At a receiver the same chipping code is employed using modulo-2 subtraction to recover the original data. The top portion of Figure 4.6 illustrates the direct sequence coding process. Prior to examining the coding process a quick review of modulo-2 operations is warranted. Table 4.2 illustrates modulo-2 addition process. Note that modulo-2 addition is similar to modulo-10 addition, with the difference being the number base used. That is, carries are discarded under any modulo system. Thus, 1+1 has a value of 0, because the carry is sent to the great bit bucket in the sky. The lower portion of Table 4.1 illustrates modulo-2 subtraction: Note that, much as with modulo-10 operations, when you subtract a number that is greater than the number you are operating on the result is a 1. Also note that modulo-2 addition and subtraction are just exclusive OR (XOR) operations.

Figure 4.6
Direct sequence uses
a chipping code to
spread each data bit.

a. Direct sequence coding

Data bits　　　1　　　　　0

chipping code	1011011	0001011
coded signal	0100100	0001011

b. Direct sequence decoding

coded signal	0100100	0001011
chipping code	1011011	1011011
Data bits	1111111	0000000

TABLE 4.2

Modulo-2
Operations

Modulo-2 Addition	Modulo-2 Subtraction
1 + 1 = 0	0 − 0 = 0
0 + 1 = 1	0 − 1 = 1
1 + 0 = 1	1 − 0 = 1
0 + 0 = 0	1 − 1 = 0

Using the Chipping Code

In Figure 4.6, the chipping code shown is a 7-bit code. In Figure 4.6a the coding of binary 1 and binary 0 data bits is shown using two different 7-bit chipping codes the better to reflect the fact that such codes vary. Each resulting coded signal is formed by modulo-2 addition of a data bit and the bits in the chipping code. In Figure 4.6b two examples of direct sequence decoding are shown. Note that the receiver must have the same chipping code as the transmitter. Also note that during the decoding process the chipping code is modulo-2 subtracted to form—or perhaps a better term—reconstruct the data bits.

The direct sequence coding process in Figure 4.6a converts each data bit into a sequence of seven-coded bit signals because a 7-bit chipping code is used. In an IEEE 802.11 wireless LAN environment, an 11-bit chipping code is employed. A receiver uses a correlation to determine how many of the received chips match the 1 bit in the 11-bit received pattern and how many match the 0 bit. The greater of the two matches then determines if the 11 bits in the received signal are interpreted as a 1 data bit or a 0 data bit. For example, if 4 of the 11 bits are recovered as 0s while 7 are recovered as 1s, the data bit is then assumed to represent a binary 1. Because the modulation process spreads the bits over a relatively large bandwidth, DSSS provides a degree of immunity to noise. That noise can represent multipath reflections or enemy jamming; since it is doubtful that our corporate wireless LAN will be jammed, we will not focus any attention on jamming.

Bandwidth Spreading

In Chapter 2 we noted how Claude Shannon's equation defined the capacity of a communications channel in terms of its bandwidth (w) in Hz, signal power (s), and noise power (n). As a refresher, the capacity of a channel in bps according to Shannon's Law is:

$$C = w\log_2 (1 + s/n)$$

Shannon's Law notes that the effect of increasing bandwidth permits the signal-to-noise ratio to be decreased without altering the capacity of the channel. In a DSSS environment the spreading of bandwidth is designed with Shannon's Law in mind. That is, if communications are spread over a wider bandwidth it becomes possible to reduce the S/N rates while keeping the bit error rate performance and resulting channel capacity at a desired level.

The process of spreading a bandwidth from a narrowband into a wideband signal is shown in Figure 4.7. At the receiver the signal is de-spread to recover the original data bits. The key parameter of interest in DSSS is the processing gain (PG), because it provides for increased system performance by requiring a high S/N ratio. Mathematically, the processing gain is:

$$PG = \frac{BW_{RF}}{BW_{INFO}}$$

where BW_{RF} represents the RF bandwidth in Hertz, and BW_{info} represents the narrow bandwidth normally required to transmit information.

Based on this equation, the processing gain for DSSS is the ratio of the chip rate to the bit rate. The larger the overall bandwidth used, the higher the processing gain for a constant data rate. Because a higher processing gain implies enhanced immunity against interference, it becomes possible for direct sequence spread spectrum to operate at a very low or even a negative S/N ratio if there is sufficient processing gain.

Figure 4.7
DSSS bandwidth spreading.

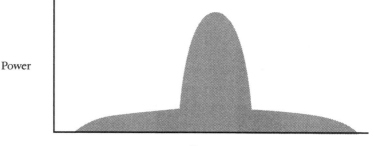

Power

Frequency

Advantages of Use

Other advantages associated with the use of direct spread spectrum systems, in addition to a low or negative S/N ratio, include its increased level of privacy and its immunity to interference. The increased level of privacy results from the fact that an RF receiver at a minimum needs to know the chipping code to recover data that would otherwise appear as noise. While the algorithms are well known, note that the word *privacy* and not the word *security* was used. If you need security you should employ encryption instead of depending on the manner by which DSSS communicates. As we previously noted, by spreading a signal over a wider bandwidth it becomes possible to overcome the effect of noise within a portion of the frequency spectrum. Unfortunately these advantages are not without some cost. Let us turn our attention to a few disadvantages associated with the use of DSSS.

Disadvantages

Two key disadvantages associated with the use of DSSS are in the areas of power and cost of ownership. Concerning power, according to studies from one vendor who makes both FHSS and DSSS wireless LAN products, the latter consumes approximately twice the power during transmission. While this results in only approximately 600mA of current in comparison to 300mA for a FHSS adapter card, it is important to note that many wireless LAN clients are mobile devices working off battery power. Thus, this can result in the necessity to provide a source of electricity for prolonged operations.

The cost of ownership between FHSS- and DSSS-compliant devices also warrants attention. While economies of scale, due to increased production, lowered the cost of products using each method, DSSS products are slightly more expensive. This tells only part of the story, however, and may only be applicable if you are considering a single access point to serve users within a geographic area. If your organization requires added capacity, however, you then need to consider both scalability and the total cost required to support your networking requirement.

A DSSS system has a limit of three nonoverlapping channels that can be obtained by collocating three access points within a geographic area or cell. In an IEEE 802.11 wireless LAN environment, DSSS can operate at either 1 or 2 Mbps. Thus, when operating at 2 Mbps the total capacity of a cell with three access points is limited to 6 Mbps. For FHSS the upper limit is more difficult to determine because there are seventy-nine possible hopping sequences. Although there is a 1/79 probability of two FHSS systems having a collision, as the number of FHSS systems increases the probability of hop collisions increases. As collisions occur, data throughput declines, resulting in FHSS with a cell capacity shaped similar to that shown in Figure 4.8. In examining Figure 4.8, note that as the number of access points within a cell rises, at first the number of resulting collisions and re-transmissions only marginally adversely affects throughput. As additional access points are deployed within a cell, however, the number of collisions and resulting retransmissions adversely affects throughput, with more frames having to be retransmitted. This explains why cell throughput and its overall capacity begins to decrease after a certain number of access points are installed. FHSS vendor literature suggests that the maximum capacity of a FHSS cell occurs with seven to nine access points that cumulatively provide approximately 8 Mbps of capacity for multiple 1 Mbps FHSS systems. (Unfortunately, I was not able to locate information concerning the cell capacity of 2 Mbps FHSS systems.)

Figure 4.8
FHSS cell capacity.

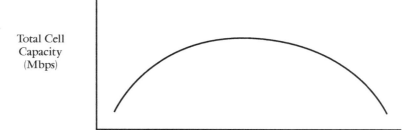

Total Cell
Capacity
(Mbps)

Number of Access Points

Coded Orthogonal Frequency Division Multiplexing

In concluding this chapter, we turn our attention to a third method used by wireless LANs as a communications system. That method is referred to as coded orthogonal frequency division multiplexing (COFDM). Although COFDM uses a wide band of frequencies it is not a spread spectrum communications method. Instead, COFDM essentially is considered a mechanism for replacing one high-speed data carrier with many lower-speed subcarriers that are transmitted in parallel to one another.

Evolution

Although COFDM represents a recently developed communications method employed in wireless LANs, prior to its adoption in the IEEE 802.11a extension to the 802.11 standard the principles behind this communications method have at least a 15-year history in the dial-up and leased-line modem world. In fact, one of the first high-speed modems I used was the Telebit Trailblazer. During the mid-1980s, this modem operated at the then-unheard-of data rate of 9600bps and was selected to enable hundreds of government investigators to use the public switched telephone network to transmit reports of their fieldwork into a central database. The key to the Trailblazer's ability to transmit at a high data rate was the fact that it used up to 256 carrier tones across the 3000Hz pass band of the voice grade channel. Data was impressed on each carrier tone using quadrature amplitude modulation (QAM), resulting in a large number of tones, each conveying a small number of bits that cumulatively provided a 9600bps transmission rate. When one modem dialed another, the handshaking process sounded like whales talking if you left the modem speaker on. During the handshaking process the sound you heard represented one modem communicating with another to determine which subcarriers were available for use and represented a training process that enabled the best subchannels to be selected for use.

If we fast forwarded about 15 years from the introduction of the Telebit Trailblazer we would note that a similar technology of multiple carriers forms the basis for the transmission scheme used by a popular type of *asymmetrical digital subscriber line* (ADSL) modem.

When used in an ADSL modem the technology is referred to as *discrete multitone* (DMT) modulation; however, this is simply another name used to describe orthogonal frequency division multiplexing. OFDM is also referred to as *multitone* or *multicarrier transmission* and is used in such broadcast systems as the *European Telecommunications Standard Institute* (ETSI) *Digital Audio Broadcasting* (DAB) and *ETSI Digital Video Broadcasting-Terrestrial* (DVB-T). Thus, COFDM's use of multiple carriers does not represent a modern innovation. Instead, it represents the application of previously tried technology into the wireless LAN environment.

Now that we have an appreciation for the evolution of multiple carrier technology, let us turn our attention to specifics and determine what the acronym COFDM really means and how the system operates.

Overview

To obtain an appreciation of what coded orthogonal frequency division multiplexing means, let us examine the components of the term. Let us first focus our attention on *frequency division multiplexing* (FDM).

FDM. In frequency division multiplexing (FDM), transmission occurs on separate channels, with channels separated by guard bands, as illustrated in Figure 4.9. The guard band simply represents unused frequency that enables a degree of frequency drift on each channel to occur without transmission on one channel adversely affecting transmission on another channel. Because multiple subchannels flow on one physical circuit we are multiplexing information transported on many channels by frequency.

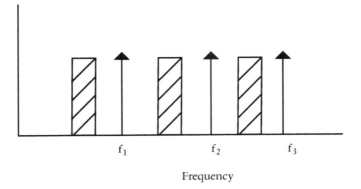

Figure 4.9
In FDM, signals are separated from one another by unused frequency referred to as a guard band.

Frequency

Legend:

fn Subchannel at Frequency fn

Guard Band

ORTHOGONAL FDM. The term *orthogonal* is used to note that signals are transmitted at right angles to one another. If you examine Figure 4.9, you note that each channel is orthogonal to the other channels. Thus, although the term orthogonal prefixes FDM it is probably a bit redundant—no pun intended. (It does, however, make it difficult for lay persons to understand communications technology and the enhancement of acronyms and mnemonics might provide some additional job security. Who said the field of data communications was dull?)

CODED OFDM. Because almost all current implementations of OFDM use some form of forward error correction (FEC) the use of FEC with OFDM resulted in the term *coded* commonly used as a prefix to OFDM, resulting in coded orthogonal frequency division multiplexing being used to define the communications method shown in Figure 4.9. Although most versions of OFDM use some form of forward error correction, the IEEE 802.11a standard relies on retransmission of frames to correct bit errors. A convolutional coder is used to insert redundant bits, which results in modulation methods that have a lower error rate. Because a convolutional coder is used, we can say that the IEEE 802.11a standard uses COFDM although a purist, from a technical perspective, might disagree. Thus, in the remainder of this section we use COFDM and OFDM synonymously with respect to the 802.11a standard.

Now that we have an appreciation of the terms that make up the name of the technology let us examine how it operates in a wireless LAN environment. In doing so we will focus our attention on its operation under the IEEE 802.11a extension to the 802.11 standard.

Operation

The use of COFDM was selected by the IEEE as a new communications method for wireless LANs when the FCC allocated 300 MHz of spectrum in the unlicensed ISM 5 GHz band. As a refresher, 200 MHz is at 5.15 GHz to 5.35 GHz, with the remaining 100 MHz of bandwidth at 5.725 GHz to 5.825 GHz. The first 100 MHz in the lower section is restricted to a maximum power output of 50 mw. The second 100 MHz has a maximum power output of 250 mw, while the third 100 MHz, which is designated for outdoor applications, can have a maximum of 1w of output power.

With three 100 MHz segments of frequency available for use, the IEEE turned to the use of OFDM as the communications method for the 802.11a extension to the 802.11 standard. Under the IEEE 802.11a standard 20 MHz of bandwidth is used to support fifty-two subchannels: forty-eight that are used to transport data and four that function as pilot channels. The pilot subcarriers function as reference signals. Those reference signals permit a receiver to disregard frequency or phase shifts because it knows the pilot frequencies and can determine the shift from the pilot. Although transmission on the carriers is orthogonal, the modulation of each carrier results in signals overlapping. This action, which is illustrated in Figure 4.10, has a positive effect on spectral efficiency because each of the carriers transporting information is positioned far enough from the others to minimize interference. As indicated in Figure 4.10, using OFDM the peak of each carrier occurs when the contribution of all other subcarriers is zero.

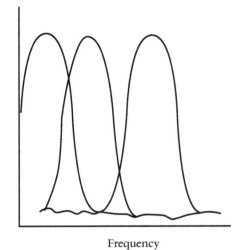

Power

Frequency

Scrambling and Coding

Prior to data bits' being modulated and placed onto a series of orthogonal carriers, the data sequence is scrambled and coded. The need for a scrambler results from the fact that any data stream can contain an arbitrary bit pattern. If the pattern contains long runs of the same value the data will not provide the receiver with a sufficient number of transitions for synchronization. Thus, the scrambling process is similar to that of a conventional modem in that the scrambler represents a feedback register that ensures the data bit stream is modified to produce enough changes from 0 to 1 (and vice versa) to enable the receiver to synchronize itself with the data stream. Of course, the receiver has its own descrambler to reproduce the original data stream. The coder used in OFDM is a convolutional coder, which adds redundancy that reduces the probability of bit errors. Next, data is interleaved as a mechanism to prevent error bursts from being input to the convolutional decoder at the receiver. The interleaved data is then mapped into data symbols for placement into the frequency domain. This mapping can occur using *binary phase shift keying* (BPSK), *quadrature phase shift keying* (QPSK), 16-QAM, or 64-

QAM. As noted in Chapter 3, OFDM modulation results in the binary data stream's being divided into groups of one, two, four, or six bits based on the modulation method used. Each group represents a point in a modem constellation pattern for a subcarrier.

Last but certainly not least, OFDM communications occurs by the placement of forty-eight data symbols and four pilots being transmitted in parallel. This transmission occurs after an inverse *fast Fourier transform* (FFT) combines the subcarriers. Figure 4.11 illustrates the previously described operations at the physical layer in a block diagram.

Figure 4.11
IEEE 802.11a
physical layer
operations.

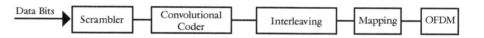

The actual data rate obtained through the use of COFDM depends on several factors. Those factors include the modulation method used, the number of bits coded per subcarrier, and the coding rate of the convolutional coder. The mixture of modulation methods and coding rates results in eight distinct bit rates. Rather than refer to the previous chapter, Table 4.3 summarizes the potential data rates obtained under the IEEE 802.11a standard. Of course, because it may be a year or two until equipment reaches the market, it will be a while until it becomes possible to verify that all modes of operation provide the indicated bit rate and data transfer rates listed in Table 4.3. In examining the entries in Table 4.3, note that the 6, 12, and 24 Mbps data rates are mandatory for all IEEE 802.11a products whereas the remaining rates are optional.

(As a refresher, the coding rate indicates the proportion of data bits to redundant bits used for error correction. Thus, a coding rate of one-half reduces the number of data bits to half of all bits transmitted. This means that the nominal data rate must be reduced by half to determine the information transfer rate.)

Advantages of Use

One of the key advantages associated with the use of OFDM is its resistance to multipath reflections. Transmission of multiple carriers at relatively low data rates results in any delayed or echoed copies of data being late by only a small fraction of a bit time. Thus, OFDM minimizes distortion caused by multipath reflections.

A second advantage of OFDM is that it provides more immunity against fading than a single carrier technique does. This greater immunity occurs because of the use of multiple carriers, where it is highly unlikely that more than a few carriers will be obstructed.

A third advantage of OFDM used in the 5 GHz band is the fact that its use of forty-eight subcarriers, when combined with 64-QAM, provides for a much higher data rate than can be obtained using other communications systems. To be fair, however, it must be recognized that OFDM under the 802.11a standard results in 16.875 MHz of bandwidth being used in a 20 MHz channel. Thus, it is actually the bandwidth availability coupled with the use of forty-eight subcarriers that provides the high data transfer capability.

Disadvantages

As in other communications methods, OFDM has certain negatives associated with its use. First, the use of multiple carriers makes this communications technique more sensitive to carrier frequency offset and sampling clock mismatch than does the use of a single carrier system. A second disadvantage associated with OFDM is that this technique results in a limited range when used under FCC power constraints in the relatively high 5GHz frequency bank: It may require a relatively large number of access points when operating an IEEE 802.11a wireless LAN to support a desired geographic area.

TABLE 4.3 IEEE 802.11a OFDM Modes of Operation

Mode	Modulation Method	Coding Rate	Bits/ Subcarrier	Coded Bits/ OFDM Symbol	Data Bits/ OFDM Symbol	Nominal Data Rate (Mbps)	Information Transfer Rate (Mbps)
1	BPSK	1/2	1	48	24	6	3.00
2	BPSK	3/4	1	48	36	9	6.75
3	QPSK	1/2	2	96	48	12	6.00
4	QPSK	3/4	2	96	72	18	13.50
5	16 QAM	1/2	4	192	96	24	12.00
6	16 QAM	3/4	4	192	144	36	27.00
7	64 QAM	3/4	6	288	216	54	40.50
8	64 QAM	2/3	6	288	192	48	32.00

Wireless LAN Hardware

In this chapter, we will become familiar with the operational characteristics of different types of wireless LAN hardware devices. As we review these characteristics, we will also examine how the devices are used in a wireless LAN environment.

These hardware devices include access points, LAN adapter cards, bridges, and routers. For those not familiar with the evolution of communications hardware products, the first device designed to perform routing was the *gateway*. This term continues to be used in TCP/IP configuration screens which require users to enter the IP address of a gateway, which represents the local router. Because of the familiarity of this term, we refer to a wireless router using the term wireless router/gateway.

Because wireless LANs represent the primary method used by small businesses and many home users to obtain multiple connections via a single cable modem or digital subscriber line (DSL) connection to the Internet, we also examine the role of *firewalls*. As we will note later in this chapter, some vendor wireless LAN products incorporate multiple functions. Thus, when applicable, we note such products and how their employment can replace multiple devices with one device. It should be noted, however, that there is a considerable range of vendor products that, while providing a common function, contain one or more features that may make such products unique. We look at several vendor products in this chapter, but it is important to note that while these are representative of similar products there may be other products that better match your actual requirements.

Wireless Access Point

A wireless access point represents a 2 port bridge that repeats data from wired network onto a wireless network and vice versa. This action enables traffic on a wired LAN to be received by mobile devices while enabling mobile device transmissions to be viewed by members of a wired LAN. Thus, an access point provides bidirectional repeater capability between a wired LAN and wireless devices within a geographic area of coverage.

Evolution

The concept behind wireless LAN access points dates back over ten years when a series of products was developed for operation in the older ISM 900 MHz frequency band. More recent access points commonly comply with the IEEE 802.11b standard, which provides support at 11 Mbps for connections to Ethernet networks operating in the 2.4G Hz frequency band. Although a wireless LAN access point provides connectivity to a 10-Mbps Ethernet LAN, in actuality the physical connection of many products is to a 10/100-Mbps Ethernet port. You can connect the wireless LAN access point to a conventional 10Base-T hub port or to a 10/100-Mbps dual-speed hub or switch port. Thus, a 10-Mbps wireless LAN access point does not adversely affect your ability to use the device with a higher speed Ethernet network, such as a Fast Ethernet LAN operating at 100 Mbps.

Equipment Connection

To illustrate the connectivity of an access point we need to use a representative vendor product. Figure 5.1 shows a view of the SMC Networks 2652w wireless access point. Note that this device is capable of being desk or wall mounted and uses a separate AC/DC converter to generate the required 9v DC power source. Because the 2652w is software configurable, it includes a reset button which, when pressed with a pencil or paper clip, restores the device to its factory default settings.

In actuality, the primary way to reset the SMC access point is through the use of a Web browser. When you use a browser you have a reset button which, when clicked, restores factory default settings. As in most sophisticated devices that could freeze due to a number of factors including electromagnet interference, improper configuration, or even a glitch in software, however, the hardware reset represents another safety method to return the state of the device to a known condition. As we continue our tour of connectors on the SMC access point, the RJ-45 connector provides for the termination of a cable for wiring the access point to a hub or switch port. While the access point is designed to be configured via a LAN using a Web browser, it also includes an RS-232 port, enabling compatibility with RS-232 devices.

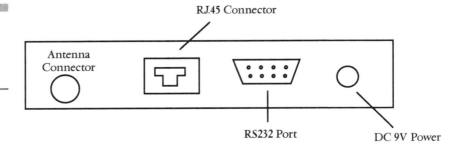

Figure 5.1
The back panel of the SMC Networks 2652w wireless access point.

The antenna connector on the rear of the SMC Networks 2652w wireless access point provides a mount for a snap-on antenna. By placing the unit at a high point within a building it is possible to obtain a line-of-sight transmission distance of approximately 1500 feet. As we note later in this chapter, a transmission distance of 1500 feet occurs at a data transmission rate of 1 Mbps and represents a best-case scenario. As you attempt to operate at a higher data rate and encounter typical office or home obstructions, such as walls and doors, the achievable transmission distance considerably decreases. Because it is normally difficult to obtain an unobstructed transmission distance anywhere near 1500 feet within a small office, industrial complex, or university campus, in all probability you will need to install multiple wireless access points if your need for wireless transmission occurs on multiple floors or is spread out over a floor with offices, building pillars, and other common obstructions. It is therefore useful to turn our attention to the use of single and multiple access points within an office environment.

Using a Single Access Point

Figure 5.2 illustrates one example of how you can use a single access point to extend the range of your organization's wired LAN. In this example the wireless access point is shown cabled to a port on a 10Base-T hub. As in any LAN device, the maximum cable distance from the hub port to the wireless LAN access point is 100 meters or approximately 328 feet. This means you have a considerable degree of latitude concerning the placement of the access point at a location that can provide a high degree of support to either fixed or mobile wireless cable computers.

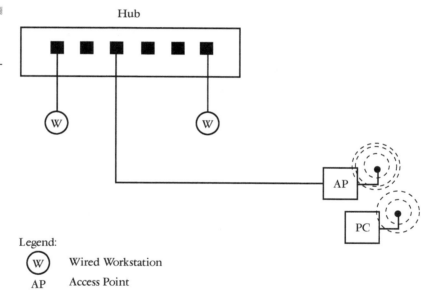

Figure 5.2
Using a single access point.

In examining Figure 5.2, note that the wireless access points I examined from several manufacturers are designed to use a standard RJ-45 connector and not a crossover cable. So that you won't be left puzzled and scratching behind your ear, make sure you cable your access point to a conventional hub or switch port and not to the "interconnection" ports on such devices that permit, for example, hubs to be interconnected to one another by including a built-in crossover on the port.

OPERATION. A wireless access point is considered to represent a repeater. Instead of repeating frames between two segments, however, the wireless access point repeats frames between the wired LAN and the radio frequency (RF) at which wireless devices operate. Thus, when a station transmits on the LAN, the access point repeats the frame at a prescribed radio frequency and in the wireless frame format, regardless of the destination of the frame. Similarly, when a wireless device transmits a frame, the access point receives the frame via the radio frequency it is set to operate on and repeats the frame onto the wired LAN.

Because an access point acts as a repeater, it is important to consider the level of utilization of the wired LAN and the potential traffic generated by wireless devices that will use the access point. You must consider the utilization of the LAN because traffic from each wireless device will be repeated by the access point onto the wired LAN. Thus, if the wired LAN is at or near saturation, the use of an access point serving a number of wireless devices will only serve to increase the level of utilization of the wired LAN, resulting in a bad problem's becoming worse. You must consider the fact that the connection of an access point to a wired LAN is similar from a traffic perspective to connecting a group of additional workstations and not a single device.

Now that we have an appreciation for the use of a single access point, let us turn our attention to the use of multiple access points.

Using Multiple Access Points

Figure 5.3 illustrates the use of two wireless LAN access points to provide an extended area of wireless access to a wired LAN. In examining Figure 5.3, note that the term *Basic Service Set* (BSS) is used to denote the communications domain for an access point operating under the IEEE 802.11 standard. When you set up a wireless LAN access point, you configure a BSS identifier. You also configure a domain identifier for those wireless LAN adapter cards that do not require roaming to the BSS ID of the access point that will serve the computer that uses the adapter card to connect with. If one or more of your wireless devices will roam between access points, you must configure two or more of your access points to create an *Extended Service Set* (ESS). By locating access points so that a continuous area of coverage is created, you provide the ability for your mobile users to roam and be serviced by an applicable wireless LAN access point. To achieve this ability to roam, both the access points and wireless network cards within a specific ESS must be configured with the same domain identifier.

Figure 5.3
Using multiple access points to support extended wireless access to a LAN.

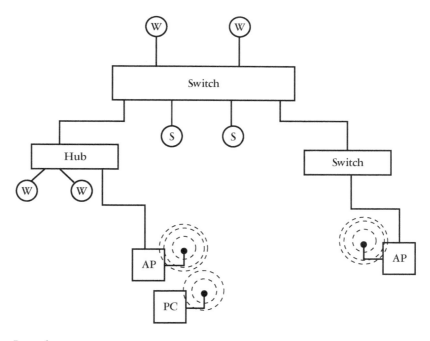

Legend:
  BSS   Basic Service Set
  ESS   Extended Service Set

(In Chapter 6 we cover the various IEEE standards in detail, and in Chapter 7 we examine the procedures involved in installing a wired LAN, including setting up a BSS and an ESS.)

Figure 5.4 illustrates the SMC Networks 2652w wireless access point. This access point is capable of supporting up to 128 users. Another SMC access point product supports up to 64 simultaneous users. Thus, you need to carefully consider the number of wireless users that may require the services of an access point before you select a specific device.

Figure 5.4
The SMC 2652 wireless access point supports up to 128 simultaneous users at data rates up to 11Mbps.
(Photograph courtesy of SMC Networks.)

In concluding our discussion of wireless LAN access points, I would be remiss if I did not discuss security. Under the IEEE 802.11 series of standards, *Wired Equivalent Privacy* (WEP) represents a mechanism that provides for the encryption of data. WEP, by default, is disabled on my devices. This lax security policy, by no means unusual, probably explains the success of two people featured in *The Wall Street Journal* April 27, 2001 article entitled "Silicon Valley's Open Secrets." In that article they discussed how they were able to drive from parking lot to parking lot in Silicon Valley, turn on their computer and, using standard equipment, locate "more than 40 corporate networks where basic security steps did not appear to have been taken." This article more than likely referred to the fact that by default WEP is disabled on most wireless LAN products. Thus, if you require security you must enable WEP on your access point and on your wireless LAN adapter cards. To do so, most products require the user to enter a phrase that functions as the security key for WEP encryption on both the access point and LAN adapter cards and enables WEP.

Concerning the enabling of WEP, most products support 64- and 128-bit encryption settings. Whereas the 128-bit setting provides a higher level of security, it requires a bit more computation—no pun intended—which makes it more suitable for use with modern computers. Unfortunately, you cannot mix and match encryption settings, so all clients of an access point must match the security setting of the access point. Because a wireless LAN access point functions as a repeater, it provides the potential to expose all of your network traffic to snoopers. Thus, the enabling of WEP should always be considered as high priority for a safe extension of your organization's wired network.

A comparison of wireless LAN access points in Table 5.1 lists some common features you may wish to evaluate. In doing so you should consider listing your requirements in the column provided and use the columns labeled Vendor A and Vendor B to match vendor products against your requirements. You can also duplicate the table and extend the last two columns to evaluate more than two vendor products.

Some of the entries in Table 5.1 warrant a degree of explanation. First, concerning network topology, an *ad hoc wireless LAN* represents a group of computers, each equipped with a wireless adapter connected via RF as an independent wireless LAN. Such computers must be configured to operate on the same radio channel and bypass an access point, which explains the omission of "ad hoc" from the table. In comparison, an *infrastructure wireless LAN* consists of an access point connected to a wired LAN and one or more wireless PC users using the access point. Wireless devices that do not require a roaming capability should be configured so that their domain identifier matches the access point domain identifier. If two or more access points are used to support roaming, both the access points and wireless adapter cards in the roaming area must be configured with the same domain identifier.

Feature	Requirement	Vendor A	Vendor B
Operation			
IEEE 801.11	_____	_____	_____
IEEE 802.11B	_____	_____	_____
Other	_____	_____	_____
Network Topology Support			
Infrastructure	_____	_____	_____
Roaming	_____	_____	_____
Configuration			
Password protected	_____	_____	_____
Via LAN	_____	_____	_____
RS-232	_____	_____	_____
USB	_____	_____	_____
OS Support	_____	_____	_____
LAN Attachment			
10Base-T	_____	_____	_____
Other	_____	_____	_____
LED Indicators			
Power	_____	_____	_____
Ethernet Link	_____	_____	_____
Ethernet Activity	_____	_____	_____
Wireless Activity	_____	_____	_____
Security			
64-bit key	_____	_____	_____
128-bit key	_____	_____	_____
Manual Settings	_____	_____	_____
Automatic Settings	_____	_____	_____

Wireless LAN Network Cards

The wireless LAN network card represents a transmitter/receiver that is installed in desktop and notebook computers. Through the use of a wireless LAN network card your desktop or notebook computer has the ability to become a node on a wireless network.

TYPES OF NETWORK CARDS. There are two primary types of wireless LAN network cards. One is fabricated as a PC card, designed for insertion into a notebook or laptop computer's PC card slot. A second type of wireless LAN network card is fabricated as an *adapter card*. This type of card is installed in the system unit of a desktop PC.

Four types of specific wireless LAN adapter cards that warrant discussion are those fabricated as PCI bus cards and those fabricated as Industrial Standard Architecture (ISA) cards, Extended ISA (EISA) cards, and as stand-alone devices with USB connectors. The use of a USB interface permits users to add wireless LAN capability to desktop computers without having to open the system expansion unit and install an adapter card in a system expansion slot.

THE SMC NETWORKS PCI CARD. Figure 5.5 illustrates the SMC Network's 2602w 11 Mbps wireless PCI card. Note that the top of the card actually represents the edge connector that is inserted into a system expansion unit slot. Also note that the right edge (shown in gray) represents the wireless LAN adapter card's antenna. This antenna protrudes from the rear of the system unit of a desktop computer, which makes the positioning of some devices difficult when you are attempting to obtain a high level of signal strength.

Although SMC Networks wireless IEEE 802.11b products are capable of supporting transmission at distances up to 1500 feet, that range is based on wireless operation at 1 Mbps in a line-of-sight environment, essentially representing a best-case scenario. To the credit of SMC Networks, their Web site provides a table of

Figure 5.5
The SMC 2602w
11-Mbps wireless
PCI card.
(Photograph courtesy
of SMC Networks.)

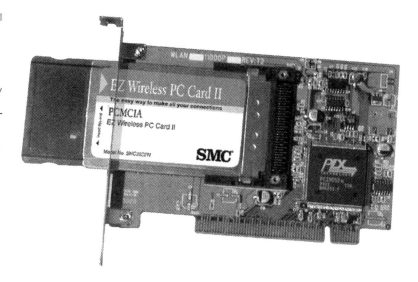

transmission distances based on operating rate and the environment of the area where their wireless products are used. Table 5.2 summarizes SMC wireless 802.11b compatible products transmission distances.

In examining the entries in Table 5.2, note that an open environment refers to a line-of-sight environment where there are no obstructions between an access point and wireless LAN network cards. A semi-open environment refers to an environment with no major obstructions, such as walls or cubicles, between an access point and wireless users. The most limited environment is the closed environment, which represents a typical home or office with floor-to-ceiling walls and door obstructions between an access point and the remote wireless LAN.

As you glance at the entries in Table 5.2, note that as the transmission rate decreases, the range of the wireless connection increases. Because the SMC Network products I used were 802.11b compatible, they automatically downspeed to a lower operating rate to provide connectivity as I moved my notebook farther from an access point.

TABLE 5.2

SMC Wireless
802.11b Maximum
Transmission
Distances

Environment	Speed and Distance Range			
	11 Mbps	5.5 Mbps	2 Mbps	1 Mbps
Open Environment	160 m	270 m	400 m	457 m
	(524 ft)	(886 ft)	(1312 ft)	(1500 ft)
Semi-Open Environment	50 m	70 m	90 m	120 m
	(164 ft)	(230 ft)	(295 ft)	(394 ft)
Closed Environment	25 m	35 m	45 m	55 m
	(82 ft)	(115 ft)	(148 ft)	(180 ft)

SMC NETWORKS WIRELESS PC CARD. A second version of a popular wireless LAN adapter card is shown in Figure 5.6. This is an SMC 2632w EZ Wireless PC Card, designed to fit into the type II PC slot included in most laptop and notebook computers. The gray area to the left of the label represents the antenna, which protrudes from your laptop or notebook computer after you install the card. Although we describe the installation and operation of the EZ Wireless PC Card in detail in Chapter 7, it should be noted that by turning your laptop or notebook to position this built-in antenna you can considerably increase the signal strength of your wireless connection. In fact, SMC Networks includes a utility program with its wireless LAN adapter cards that you can easily invoke to determine the signal strength of your wireless connection. By displaying the signal strength as you position your notebook or laptop computer, you can determine an optimum position to enhance the transmission between your computer and an access point.

In concluding our examination of wireless LAN network cards, Table 5.3 provides a list of features you may wish to consider when acquiring this type of wireless LAN product. Muach as with to the table provided for evaluating access points, you can denote your requirements and compare and contrast different vendor products by completing the table.

Figure 5.6
The SMC 2632w EZ Connect Wireless LAN card is designed for insertion into a Type II PC card slot. (Photograph courtesy of SMC Networks.)

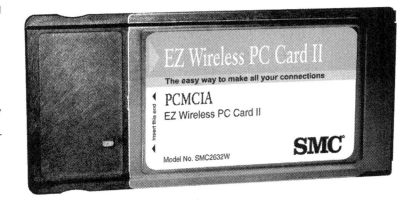

TABLE 5.3

Wireless LAN Card Features to Consider

Feature	Requirement	Vendor A	Vendor B
Operational Compatibility			
IEEE 801.11	_____	_____	_____
IEEE 801.11B	_____	_____	_____
Other	_____	_____	_____
Hardware Form			
ISA Adapter	_____	_____	_____
EISA Adapter	_____	_____	_____
PCI Adapter	_____	_____	_____
PCI Card	_____	_____	_____
Driver Support			
Windows Version	_____	_____	_____
Macintosh	_____	_____	_____
Other	_____	_____	_____
Configuration Utility			
Uni-quality Monitor	_____	_____	_____
Configuration Parameters	_____	_____	_____
Encryption Control	_____	_____	_____

In examining the entries in Table 5.3, we will defer a description of common configuration parameters until we review IEEE 802.11 network terminology in the next chapter. It should be mentioned, however, that most products I configured are designed to accept default settings without requiring a user to change any configuration settings. Unfortunately, this also means that for most products, accepting a default setting results in disabled encryption. At a minimum you should configure encryption on your adapter cards to match this feature on your access points.

Wireless Bridges

A wireless bridge represents a device that is designed to interconnect two conventional wired LANs via wireless transmission. Most wired bridges only support a limited cabling distance. Thus, if your organization needs to interconnect two geographically separated LAN segments, the most common method employed is to obtain a transmission line and a pair of conventional routers. This can represent a costly method of interconnection, however, and some organizations turn instead to the use of wireless bridges.

RATIONALE FOR USE. To understand the employment of a wireless bridge, consider a bridge designed to interconnect two 10Base-T networks as illustrated in Figure 5.7. Because the maximum cabling distance from an Ethernet adapter card to a hub port is 100 meters, a conventional wired bridge is limited to interconnecting network segments located within 200 meters of one another. Although this may be a sufficient distance for many intrabuilding applications, what happens if the networks to be interconnected are located farther apart from one another or even in geographically separated buildings? While you could install a leased line between the geographically separated locations and acquire a pair of remote bridges or routers, doing so would result in a monthly recurring cost for the leased line that would make your local telephone company very happy. Even if you attempt to interconnect two LANs within the same building, you may have to install wiring and a conduit that can represent a significant one-time expense. As an alternative to either of these scenarios, you can consider the use of a pair of wireless LAN bridges.

Ethernet Hub

Figure 5.7
The use of a conventional bridge to connect two Ethernet LANs limits the inter-LAN distance to 200 meters.

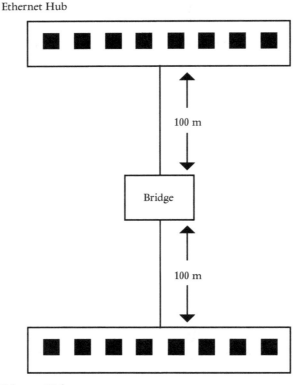

Ethernet Hub

OPERATION. A wireless LAN bridge is similar in many ways to its wired cousin. That is, it operates by following the rule of 3Fs— flooding, filtering, and forwarding. For readers not familiar with the manner in which conventional bridges operate, we first turn our attention to this topic. Once this is accomplished we will turn our attention to the manner in which a wireless LAN bridge operates as well as describe and discuss its capabilities and constraints.

OVERVIEW. A bridge operates at the Data Link layer, using the source address in frames to construct a *port address table*. This table is used to determine if frames should be forwarded at a specific port, based on the destination address in a frame. If the destination address in a frame does not have a corresponding entry in the port address table, the bridge floods the frame. During the flooding process the frame is output onto all ports other than the port on which the frame was received.

Figure 5.8 illustrates the fields in an Ethernet frame. In examining these, note that the destination and source address fields are each six bytes in length and represent the recipient and originator of the frame. As indicated in the lower portion of Figure 5.8, both destination and source addresses have two subfields. One subfield, referred to as the Vendor-ID field, is assigned to vendors by the IEEE. The Vendor-ID is three bytes in length and defines the manufacturer of an Ethernet network adapter or interface. The Manufacturer Number subfield is controlled by the vendor and identifies each Ethernet network adapter or interface that it produces. When a vendor uses all available numbers in the three-byte Manufacturer Number subfield the vendor then requests the IEEE for another Vendor-ID three-byte code and resumes manufacturing network adapters or interfaces. Thus, each Ethernet network adapter or interface has a unique address.

Figure 5.8
The fields of an Ethernet frame include six-byte destination and source address fields that are each subdivided into three-byte Vendor ID and Manufacturer Number subfields.

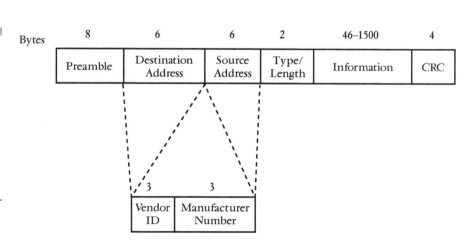

PORT ADDRESS TABLE CONSTRUCTION. To illustrate how a wired bridge operates, we need an example, but, because I don't like to work with 48-bit addresses, for illustrative purposes I will use the letters of the alphabet.

Figure 5.9 illustrates the connection of three LAN segments to a wired bridge. For simplicity we assume workstations with addresses A and B are connected to segment 1, which is connected to port 1 on the bridge; workstations with addresses C and D are connected to segment 2, which is connected to port 2 on the bridge; and workstations with addresses E and F are connected to segment 3, which is connected to port 3 on the bridge.

Figure 5.9
Bridges construct entries in their port address table by examining the source address of frames appearing on different ports.

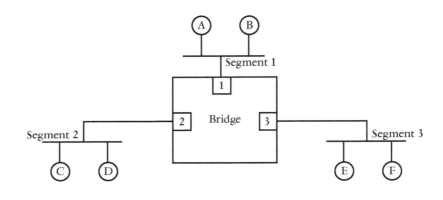

Operation		Port/Address	
1. Power-on	1.	—	—
2. A transmits to C	2.	1	A
3. C transmists to A	3.	2	C
4. B transmits to A	4.	1	B

To understand how the port address table is constructed and used, assume the bridge was just pulled out of its box, connected to each segment, and powered on. Thus the first entry in the table (located in the lower portion of Figure 5.9) indicates that there are no entries in the port address table when the bridge is first powered on. Next, let us assume the workstation whose source address is A transmits a frame to the workstation whose source address is C. When this action occurs the frame flows to all stations connected to segment 1, including port 1 on the bridge. The bridge examines the frame, which will be similar to that previously shown in Figure 5.8, noting both the destination and source addresses in the frame. The bridge then examines its port address table to determine the port onto which the frame should be forwarded by attempting to match the destination address in the frame to an address in its port address table. Because there are no entries in the port address table at this particular time, the bridge obviously cannot match the destination address in the frame to any address in the port address table. Thus, the bridge floods the frame out of ports 2 and 3. In addition, the bridge notes that the source address of A was received on port 1 and enters that association into its port address table.

For a second example of how the port address table is constructed, assume that the workstation whose address is C transmits a frame to the workstation whose address is A. The bridge receives the frame on port 2 and checks the entries in its port address table against the destination address in the frame. The bridge notes that the destination address is on port 1 and forwards the frame out of port 1 toward its destination.

The bridge also uses the source address in the frame. In doing so it compares the source address against the current entries in the port address table. Because address C is not in the port address table, the bridge adds that address to the table and associates it with the port on which the frame containing the address was received. Thus, the bridge adds address C and its association with port 2 to the table.

Next, let us turn our attention to an example of filtering, to complete our review of the three Fs as we examine the use of a bridge's port address table.

For our third example, assume that the workstation with address B transmits a frame to the workstation A. The bridge receives this frame on port 1 and checks the destination address against entries in the port address table. Because address A and its associated port 1 is in the table, the bridge prepares to forward the frame onto port 1. Because the port on which the frame was received matches the forwarding port, any forwarding would duplicate the frame on the segment. Therefore, the bridge filters the frame into the great bit bucket in the sky. The bridge also notes that the source address in the frame (B) is not currently in the port address table. Therefore, the bridge adds address B and its association with port 1 to its port address table.

LIMITATIONS.   This example of the creation of the port address table of a bridge, while accurate, avoided mention of two limitations associated with the use of the table and how those limitations are overcome. The two key interrelated port address table limitations are the maximum number of entries permitted in the table and the average search time required to locate an entry.

Because there can be thousands of workstations behind the port of a bridge, each connected to segments behind the bridge, it is possible for port address table entries to rapidly expand. As the entries in the port address table expand, the average time required to match a destination address against an entry in the table obviously increases. To prevent search times from adversely affecting the flow of frames, the number of possible entries in the port address table is limited. To ensure that an upper limit on port-address table entries does not result in the stagnation of the table, a *time stamp* is added to each entry as it is placed in the table. By periodically examining the time stamp associated with each entry, the bridge can purge old entries to make room for new ones. While we did not show a time stamp for each entry in the lower portion of Figure 5.9, as a famous radio announcer would say, "now we know the rest of the story."

WIRELESS OPERATION.   In a wireless LAN environment a bridge has two ports associated with its wired cousin. One port is cabled to a wired LAN, whereas the second port can be considered to represent the antenna of the device, which provides a radio frequency communications capability.

Figure 5.10 illustrates the use of two wireless LAN bridges to interconnect two wired networks. In examining Figure 5.10, note that the antenna of a wireless LAN bridge may be capable of functioning as multiple ports, thus providing the ability to communicate with multiple wireless LAN bridges. The ability to transmit to multiple wireless LAN bridges as well as other wireless LAN bridge features is examined next.

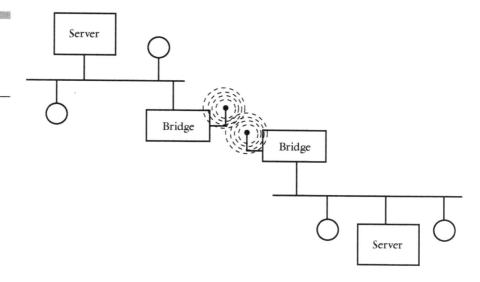

Figure 5.10
Using wireless
LAN bridges to
interconnect two
wired LANs.

FEATURES. One of the possible functions included in a wireless LAN bridge is the ability to support multiple remote bridges. Although this can be an important consideration for many organizations, it is only one of many wireless LAN bridge features that warrant attention. Table 5.4 lists a number of such features you may wish to consider when examining the potential use of wireless LAN bridges. This listing of features is provided in the form of a table to facilitate matching organizational requirements against specific vendor products.

Concerning the antenna entry in Table 5.4, some products designed for indoor use include an integral antenna. In comparison, products designed for outdoor use may have a separate antenna, which can be mounted outside, and is connected via a cable to a bridge installed inside a building.

The actual data rate supported by a wireless LAN bridge is based on the transmission method supported. While more recent manufactured bridges are beginning to support the IEEE 802.11b standard for transmission at 11 Mbps, other products support data rates from 1 to 100 Mbps, with the former requiring a large buffer because a wired LAN operates at a minimum of 10 Mbps.

TABLE 5.4

Wireless LAN Bridge Selection Features

Feature	Requirement	Vendor A	Vendor B
Antenna	_____	_____	_____
Data Rate (maximum)	_____	_____	_____
Network Interface			
RJ-45	_____	_____	_____
BNC	_____	_____	_____
Fiber	_____	_____	_____
Other	_____	_____	_____
Operational Mode			
Point-to-Point	_____	_____	_____
Point-to-Multipoint	_____	_____	_____
Maximum Number of Addresses	_____	_____	_____
Operational Range (Miles)	_____	_____	_____
Transmission Band			
2.4 GHz	_____	_____	_____
5.8 GHz	_____	_____	_____
Other	_____	_____	_____
Transmission Method			
IEEE 802.11	_____	_____	_____
IEEE 802.11b	_____	_____	_____
Proprietary	_____	_____	_____
Utilization			
Indoor	_____	_____	_____
Outdoor	_____	_____	_____

The operational mode of a wireless LAN bridge includes both the ability to transmit to one or multiple remote wireless LAN bridges and the maximum number of addresses a bridge can learn and store. The operational range of a wireless LAN bridge depends on several factors, including the type of antenna used by the bridge, its operational frequency, and its output power. Bridges that operate in the unlicensed ISM bands must comply with FCC power regulations, which results in a maximum transmission distance of approximately 10 to 20 miles for IEEE 802.11b-compatible devices. This range is based on a line-of-sight transmission, however, with no obstructions between transmitter and receiver—essentially a best-case condition.

One item not included in Table 5.4 but which warrants discussion is the use of *antenna diversity*. When transmission occurs through the use of antennas mounted on top of buildings in close proximity to one another, without obstructions between buildings, multipath propagation will more than likely be negligible and a single antenna is sufficient. For situations where multipath propagation occurs, however, some products can be obtained with dual antennas that enable the wireless LAN bridge receiver to employ *space diversity* data reception.

Figure 5.11 illustrates a BreezeNet DS.11 wireless outdoor bridge system. This system consists of two components. The shorter device shown in the left portion of Figure 5.11 represents the wireless base unit, which is installed indoors. One RJ-11 plug provides a connection to a 10Base-T LAN while the second plug provides a connection to the outdoor unit, the wireless remote bridge shown in the right portion of the figure. The antenna, which is not shown, connects to the remote bridge. Thus, this wireless LAN bridge system actually consists of three distinct hardware devices and a series of cables that are used to interconnect the units and to connect the base unit to a wired LAN.

Wireless Router/Gateway

Many small businesses and home offices obtain connectivity to the Internet via a digital subscriber line (DSL) or cable modem connection. Although you can install a wired hub and run cables to other locations to support connectivity between multiple users, you cannot share the Internet connection unless you install a separate gateway or router. That device more often than not must be capable of performing *network address translation* (NAT), because most DSL and cable model service providers use the *Dynamic Host Configuration Protocol* (DHCP) to temporarily lease the subscriber a single IP address. Thus, any additional computers behind the DSL or cable modem must obtain the capability to become addressable if such computers need to access the Internet. As you might expect, the key to this addressing functionality is obtained by the NAT capability of a router or gateway. Because you more than likely want a maximum degree of flexibility, you should consider the use of a wireless router or gateway that can service both wired and wireless connections.

OVERVIEW. A wireless router or gateway is a device that allows multiple computers to share wireless access to the Internet or another network via a single wired connection, such as a DSL or cable modem connection. While we normally consider the Internet connection of a router or gateway to represent a high-speed transmission facility, it should be noted that some products also include a built-in V.90 56Kbps analog modem. Such products can be used to access the Internet at a DSL or cable modem-operating rate as well as provide shared access via an analog modem. Although most readers probably prefer the use of wireless transmission to provide shared access to a high-speed Internet connection, the built-in modem can be used as a backup mechanism in the event your primary method of high-speed access fails.

Another feature of wireless routers or gateways that deserves mention is the manner in which they service clients. The wireless router or gateway may be limited to supporting wireless connections or it may support both wired and wireless connections.

Figure 5.12 illustrates the operation of two types of wireless router/gateway products. In the top portion of Figure 5.12 a wireless router/gateway that is limited to supporting only wireless stations is shown. In the lower portion of Figure 5.12 a wireless router/gateway that supports both wired and wireless connectivity is shown. For either type of product you normally first cable the device to your DSL or cable modem. You then connect a PC to the wireless router or gateway to configure the device.

A wireless router or gateway that only supports wireless communications will normally include either a USB or an RS-232 configuration port. A wireless router or gateway that supports both wired and wireless communications typically includes a wired hub or mini-LAN switch embedded in the device. This enables the configuration process to be performed through the use of a browser, once you cable a PC to a port on the hub and enter the address of the device.

The lower portion of Figure 5.12 illustrates the use of a wireless broadband router or gateway to provide both wired and wireless access to the Internet. In this example, the wireless router or gateway is shown supporting three wired LAN ports. One port connects to the DSL or cable modem while the remaining two LAN ports permit individual PCs to be connected via a cable from their Ethernet adapters. For both versions of wireless router/gateway devices shown in Figure 5.12, the wireless transmission provides support for other devices that have shared access to the Internet.

Figure 5.12
There are two basic types of wireless router/gateways—those that support only wireless communications and those that add support for wired LAN connectivity.

A. Using a wireless router/gateway limited to supporting wireless stations.

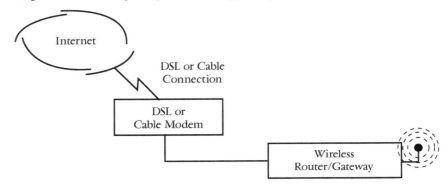

B. Using a wireless router/gateway that supports both wired and wireless stations.

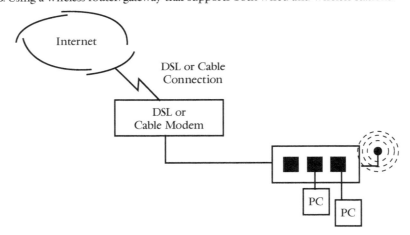

NETWORK ADDRESS TRANSLATION (NAT). Several devices that I examined employed RFC 1918 addresses as a mechanism to extend an IP addressing capability behind a DSL or cable modem to multiple devices. As a refresher for readers who are not familiar with RFC 1918, that Request For Comment defines three blocks of IP addresses for private internets. Thus, an organization can assign RFC 1918 addresses to devices behind a router or gateway and access the Internet as long as a mechanism is available to translate RFC 1918 addresses into one or more addresses that can be used on the Internet.

The reason RFC 1918 addresses cannot be used on the Internet is because such addresses are reserved for private use; they can be used by multiple organizations. Because each address on the Internet must be unique you cannot directly connect devices with RFC 1918 addresses to the Internet; this would confuse routers about where such addresses actually reside. Instead, such addresses are translated via NAT into unique IP addressing.

As mentioned earlier, most DSL and cable modem subscribers have only one IP address. In addition, that IP address is normally leased via a DHCP server. Thus, the router or gateway must provide a mechanism to translate RFC 1918 addresses to the single address your ISP provides. To accomplish this, some products I examined translated RFC 1918 addresses to high port numbers, allowing the single leased IP address initially associated with the computer behind the DSL or cable modem to support literally hundreds of distinct computers.

REPRESENTATIVE PRODUCTS. To obtain an appreciation of some of the features of wireless routers and gateways you may wish to evaluate, we turn our attention to two products. First, we look at the SMC Networks 7004 WBR, better known as the Barricade wireless broadband router. Once we examine its features, we then turn our attention to the Agere Systems' Residential Gateway.

SMC BARRICADE. The SMC Networks Barricade wireless broadband router is illustrated in Figure 5.13. This router/gateway product represents one of the most sophisticated low-cost wireless LAN products I encountered after examining literally hundreds of products. In addition to functioning as an access point to provide wireless connectivity to a wired LAN, which can consist of a single connection to a DSL or cable modem, the SMC Networks Barricade is a three-port 10/100 Mbps LAN switch, firewall, and network address translation device slightly bigger than four packages of cigarettes.

Note four ports shown in the center portion of Figure 5.13; the port on the left is labeled "WAN." You can connect a cable modem or DSL modem to this port. The three ports to the right of the WAN port represent three 10/100Base-T Ethernet switch ports. You can cable a conventional Ethernet hub or PC with an Ethernet adapter card to any of those ports.

If you examine the rear of the SMC Networks Barricade wireless wideband router, you note that this device has dual antennas. This permits the best signal to be selected and enhances the reception of data. Up to 128 wireless devices can be supported by the Barricade.

In addition to its wired and wireless LAN support, the Barricade includes a built-in print server capability. This enables wireless and wired LAN users to share the use of a common printer connected to the wireless broadband router.

Figure 5.13
The SMC wireless broadband router includes an access point, print server, firewall, dual antennas, and the ability to support both wired and wireless shared Internet access through its network address translation capability.

The SMC Networks Barricade includes network address translation (NAT) capability. The Barricade uses the RFC 1918 IP address of 192.168.123.254 by default and was configured by me to support both the wired and wireless stations that will be described in considerable detail in Chapter 7. The SMC Networks Barricade performs NAT transparently to users, permitting one IP address provided by an Internet Service Provider to be shared among many wired and wireless users.

ORINOCO RESIDENTIAL GATEWAY.   Until early 2001, the Orinoco Residential Gateway, was a product of Lucent Technologies. In early 2001 the Microelectronics Group of Lucent Technologies, which is responsible for wireless LAN products, was spun off as an independent company known as Agere Systems. Thus, the Orinoco Residential Gateway is now a product of Agere Systems.

Figure 5.14 illustrates an Agere Systems Orinoco starter kit. This kit includes an Orinoco RG1000 residential gateway, shown in the right portion of Figure 5.14, and a Orinoco World PC Card, shown in the left portion of the figure.

The RG1000 residential gateway includes three built-in interfaces—an RJ-45 socket for a 10Base-T Ethernet interface, an RJ-11 socket to provide telephone connectivity to a built-in v.90 56 Kbps modem, and a wireless IEEE 802.11b air interface. Thus, you can use the RG1000 to construct a wireless network in your home or small business and obtain wireless access to the Internet via a telephone line using the residential gateway's built-in 56 Kbps modem or via the Ethernet interface to a DSL or cable modem. When used as a wireless LAN device the residential gateway forwards data from one wireless computer to another.

Similar to the SMC Networks wireless broadband router, the Orinoco residential gateway includes network address translation capability, which permits multiple wireless clients to access the Internet via a common 56-Kbps analog modem, DSL, or cable modem connection. The RG1000 can be mounted on a wall or on a desk and has a maximum transmission range of 1750 feet when operating at 1 Mbps in an open environment. At its maximum data rate of 11 Mbps the maximum range is reduced to 525 feet; however, when used in a typical home or office environment a more practical transmission distance is 150 feet.

In early 2001 Agere implemented DHCP and added support for *Point-to-Point Protocol over Ethernet* (PPPoE) as a feature for the residential gateway. These two additions, when added to NAT, enable the residential gateway to be used with a variety of broadband connections. Two additional features included in the Orinoco residential gateway include built-in *Voice-over-IP* (VoIP) support for Spectralink wireless voice handsets and Telnet support. The latter permits network managers to maintain their wireless network remotely.

Figure 5.14
The Agere Systems Orinoco starter kit includes a residential gateway and a World PC card.

FEATURES. In concluding our examination of wireless router/gateway products, we note the general features you may wish to consider when shopping for these products. Table 5.5 shows router/gateway features in a format similar to other equipment evaluation tables presented earlier in this chapter.

TABLE 5.5

Wireless Router/Gateway Selection Features

Feature	Requirement	Vendor A	Vendor B
Internet Access			
Modem	___	___	___
DSL	___	___	___
Cable Modem	___	___	___
NAT Support	___	___	___
DHCP Support	___	___	___
Wired Connections			
Ethernet Ports	___	___	___
Other	___	___	___
Wireless Connections			
Clients Supported	___	___	___
64-bit Key Security	___	___	___
128-bit Key Security	___	___	___
Wireless Transmission Distance			
Open Environment	___	___	___
Closed Environment	___	___	___
Management			
Telnet	___	___	___
Web Browser	___	___	___
RS-232	___	___	___

IEEE Wireless LAN Standards

In previous chapters we became acquainted with wireless technology, including different modulation techniques used by wireless LANs and the role of different generic types of hardware products. Using the previous information as a base, we now turn our attention to a troika of existing Institute of Electrical and Electronic Engineers (IEEE) wireless LAN standards. Before we discuss those standards, a few words concerning the designators used to denote each standard are in order.

The first IEEE wireless LAN standard is designated as the *802.11 standard*. This standard defines LAN operating rates of 1 and 2 Mbps using infrared, frequency hopping spread spectrum (FHSS), and direct sequence spread spectrum (DSSS) technology. Although you might logically expect the second IEEE wireless LAN standard to add a suffix of 'i' or 'a' to the 802.11 designator, in actuality the second standard to be defined is designated as an extension to the IEEE 802.11 standard and is referred to as the *802.11b standard*. This standard defines the use of DSSS at data rates of 1, 2, 5.5 and 11 Mbps. The third wireless LAN standard also represents an extension to the IEEE 802.11 standard and has the designator *802.11a*. This standard supports data rates up to 54 Mbps; however, as this book was written products supporting the standard were in beta testing. Now that we have an appreciation for the basic meaning of the three IEEE standards let us turn our attention to specifics.

The 802.11 Standards

The final draft of the first IEEE wireless LAN standard was approved on June 26, 1997. In actuality, the efforts of the IEEE in the area of wireless LANs date back to 1990, when the organization formed its 802.11 Wireless Local Area Network Standards Working Group. The IEEE 802.11 working group was initially assigned the task of developing a global standard for wireless LANs operating in the 2.4 GHz unlicensed frequency band and in infrared at data transmission rates of 1 and 2 Mbps. After a series of draft documents were developed, the initial standard was finalized in mid-1997. This standard, which is officially titled "IEEE Standard for Wireless LAN Medium Access Control (MAC) and Physical Layer (PHY) Specifications" defines over-the-air protocols that can support networking in a local area.

Overview

As with other IEEE 802-based standards including Ethernet (802.3) and Token Ring (802.5), the objective of the 802.11 standard is to define a mechanism for the delivery of *MAC Service Data Units* (MSDUs) between peer *Logical Link Controls* (LLCs). As a refresher for those readers not familiar with LAN standards, the IEEE subdivided the Data Link layer in the International Standards Organization (ISO) protocol stack. The result of this subdivision, which is shown in Figure 6.1, split the Data Link layer into a MAC layer and a LLC layer. The Medium Access Control sublayer is responsible for controlling access to the network. In comparison, the Logical Link Control sublayer provides a mechanism for generating and interpreting commands as well as performing recovery operations in the event of errors. In doing so, the LLC provides a link between network layer protocols and media access control.

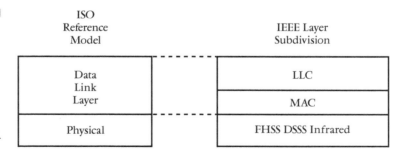

Figure 6.1
All IEEE 802-based standards subdivide the ISO's Data Link layer into a media access control (MAC) and logical link control (LLC) sublayer.

The subdivision of the data link layer into MAC and LLC sublayers permits a common control for different media access. Similar to the subdivision of the Data Link layer the IEEE also, when necessary, subdivides the Physical layer. This subdivision makes it possible to support different wireless media, such as different types of radio frequency (RF) signaling and infrared transmission. Figure 6.2 illustrates the manner in which the IEEE subdivided the ISO's Physical layer. The *Physical Layer Convergence Procedure* (PLCP) sublayer defines the manner by which MAC sublayer protocol data units (MPDUs) are mapped into a frame format suitable for the *Physical Medium Dependent* (PMD) sublayer. The PLCP can also perform carrier sensing for the MAC sublayer. In comparison, the PMD sublayer supports the appropriate over-the-air medium to be used. In doing so the PMD defines the method for transmitting and receiving data through the medium, including data coding and modulation.

Figure 6.2
The IEEE subdivided the Physical layer to enable a common Physical Layer Convergence Procedure (PLCP) to support different data transmission methods defined by the Physical Medium Dependent (PMD) sublayer.

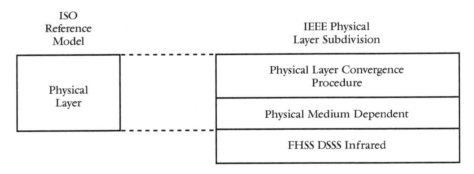

Topology

The IEEE 802.11 standard supports two types of topologies, a basic service set (BSS) and an extended service set (ESS).

The basic service set represents the basic building block of an IEEE 802.11 wireless LAN. The BSS consists of two or more mobile nodes that can communicate among one another, in effect supporting peer-to-peer communications. Each node is also referred to as a *station* and is free to move within the BSS. However, a station can no longer communicate with other stations if it leaves the BSS.

Figure 6.3 illustrates three stations within a BSS. Note that the stations are limited to communicating with one another and do not have the ability to communicate with stations on a wired LAN. Each BSS has an identification, which normally corresponds to the MAC address of the network interface card. Referred to as the *BSSID*, this identifier is six bytes in length, which is the same length as the identifiers used on wired LANs. The area of wireless coverage within which members of a BSS can communicate is referred to as a *Basic Service Area* (BSA).

THE ESS.   Two or more BSSs are connected to form an extended service set (ESS) via the use of two or more access points. When access points are employed to form an ESS, the infrastructure that connects the BSSs together is referred to as a *distribution system*.

Figure 6.3
A peer-to-peer
wireless LAN.

THE DISTRIBUTION SYSTEM.   Figure 6.4 illustrates the relationship between three BSSs, an ESS, three access points, and a distribution system. Note that the distribution system can represent an existing LAN infrastructure or a twisted-pair wire that simply interconnects two access points. In fact, the IEEE 802.11 standard does not place any constraints on the composition of the distribution system. Thus, it could represent an 802.3 Ethernet, 802.5 Token-Ring network, or even a non-IEEE LAN.

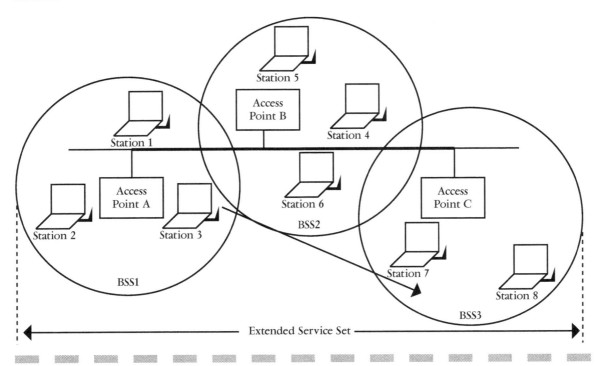

Figure 6.4 A distribution system that interconnects access points also interconnects basic service sets to form an extended service set.

ACCESS POINTS. Each access point has a unique address and represents a wired-to-wireless bridge, which provides an interface to the distribution system for stations located within different BSSs. Similar to the BSSs, the ESS has an identification. That identification is referred to as the *ESSID*. By defining a common ESSID you enable a mobile station to roam from one BSS to another while being serviced by different access points.

In the example illustrated in Figure 6.4, station 3, which is a member of BSS 1, is shown roaming into BSS 3. Because a common ESSD is defined, a mobile station in any BSS can roam into another BSS.

While an access point provides a wireless-to-wired LAN connection point, it also functions as a radio relay to mobile or fixed wireless stations located within a geographic area. In this situation, stations are not required to communicate with one another.

Instead, if one station needs to transmit to another station it does so through the services of an access point. Although this doubles the use of bandwidth it reduces the required level of performance of the wireless link. This is because stations do not need to have enough power and good antenna positioning to reach all other stations within a geographic area. Instead, they only need the ability to communicate with the access point.

CLIENT SUPPORT. The actual number of clients that can be supported by an access point depends on the transmission activity of clients serviced by that access point. Figure 6.5 illustrates the use of an access point to extend the range of a wired LAN. In general, an access point can normally accommodate between ten and twenty client stations when clients require a moderate to high level of LAN access. If clients require a low level of wired LAN access, it may be possible to support up to fifty and possibly additional clients through the use of a single access point. Of course, the distribution of clients within the geographic area served by the access point, as well as the presence or absence of obstructions to the wireless signal, also governs the client support

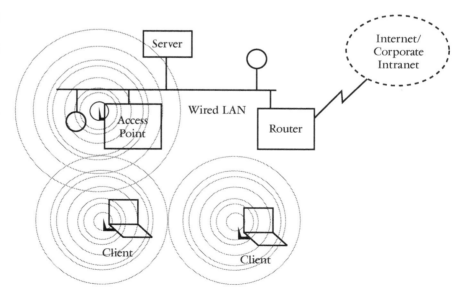

Figure 6.5
Using an access point to extend the range of a wired LAN by allowing wireless client access.

of an access point. Because of this, the use of access points serving more than a dozen or so clients commonly represents a trial-and-error procedure.

TRANSMISSION DISTANCE. Because the transmission power of wireless LANs is significantly less than that of mobile phones, a wireless LAN station can only transmit at a distance that is a fraction of that obtainable by a cellular phone. Although the actual transmission distance obtained depends on the method of transmission used and obstructions between clients and access points, most access points are realistically limited to supporting a transmission radius of approximately 100 to 200 feet within a typical office or home environment.

Portals

Another potential component of an 802.11 wireless LAN that deserves mention is a *portal*. A portal represents a translation bridge, because it provides the interconnection capability between an 802.11 wireless LAN and a wired LAN. Because most vendors include the access point and portal as a single physical entity, the portal is rarely marketed as a distinct hardware product.

The Physical Layer

Under the IEEE 802.11 standard, three physical layers are specified. Two RF physical layers are defined for operation in the 2.4-GHz ISM frequency band, while one infrared layer is defined. Although you can use any physical layer for communications within a BSS, to interoperate all devices within a BSS, each device must conform to the same physical layer mode of operation.

Both RF methods are defined for operation in the 2.4-GHz frequency band, typically occupying 83 MHz of bandwidth from 2.400 GHz to 2.483 GHz. The selection of this frequency, while certified for use by over 40 countries, can vary with respect to permissible power. For example, the FCC in the United States limits radiated antenna power to 1w. In Europe the radiated RF power level is limited to 10 mw per 1 MHz, whereas in Japan it is limited to 10 mw. In addition, there are differences in the allocation of certain frequencies for RF Physical layer use between countries that precludes the ability to easily transfer equipment between the offices of a multinational company. Now that we have an overview of the physical layer, let us focus on specifics by reviewing how each physical layer is implemented.

FREQUENCY HOPPING SPREAD SPECTRUM. Under the IEEE 802.11 standard, frequency hopping spread spectrum communications represents one of three physical layers supported by the standard. The IEEE 802.11 frequency hopping physical layer uses seventy-nine nonoverlapping frequency channels, with each channel having a 1 MHz channel spacing. This enables up to twenty-six collocated networks to operate, which can provide a reasonably high aggregate throughput.

FREQUENCY ALLOCATION AND HOPPING CHANNELS. Although the standard specifies seventy-nine nonoverlapped frequency channels, the actual number of channels used and their power and frequency assignment depends on the regulatory assignment of frequency usage in a particular country.

HOPPING CHANNELS. Once the number of frequency channels is defined, those frequency channels can be used to construct a set of hopping patterns that minimizes the probability of one BSS, operating on the same frequency channel at the same time as another BSS, thereby minimizing interference while enhancing the support for a larger number of wireless clients within a geographic area by supporting multiple access points. In the United States there are three sets of hopping patterns defined for FHSS, with twenty-six hopping patterns in each set, resulting in a total of 3 × 26 or seventy-eight hopping patterns. Thus, within a common geographic area it is possible to collocate three access points within close proximity of one another. Because different regulatory bodies restrict the number of frequency channels available for use, this also affects the number of hopping patterns per channel as well as the total number of hopping patterns. Table 6.1 summarizes the allocation of frequency channels, number of sets of hopping patterns, number of hopping patterns in a hopping set, and total number of hopping patterns for North America and four other locations around the globe.

TABLE 6.1 FHSS Frequency Channel and Hopping Patterns

Geographic Area/ Country	Minimum Number of Frequency Channels	Maximum Number of Frequency Channels	Number of Sets of Hopping Patterns	Number of Hopping Patterns Per Set	Total Number of Hopping Patterns
North America	75	79	3	26	78
Western Europe Except Spain, France	20	79	3	26	78
France	20	35	3	11	33
Spain	20	27	3	9	27
Japan	10	23	3	4	12

Modulation

Under the 802.11 standard the use of frequency shift keying (FSK) is used for FHSS because of its low cost and easy operation. Actually, there are two versions of FSK specified, each more formally referred to as *Gaussian-shaped FSK* (GFSK). Using GFSK, which operates at 1M symbols/s, nonreturn to zero (NRZ) data is filtered via the use of a low-pass Gaussian filter and then the filtered result is used to frequency-modulate a carrier. To provide a 1 Mbps operating rate, which is mandatory, a two-level GFSK modulation method is used, with binary 1s and 0s modulated into one of two frequencies. To support the optional 2 Mbps data rate a four-level GFSK modulation method is employed, with pairs of bits modulated using one of four frequencies.

Frame Format

At the physical layer, FHSS transmits data using a predefined frame format. Figure 6.6 illustrates the 802.11 frequency hopping spread spectrum frame format.

Figure 6.6
IEEE 802.11
frequency hopping
spread spectrum
frame format.

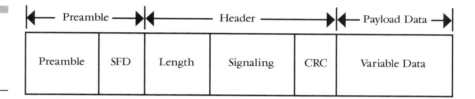

Preamble field is an 80-bit sequence that has the pattern 0101 . . 0101.

SFD field is the Start of Frame Delimiter that has the pattern 0000 1100 1011 1101.

Length field is 12 bits in length.

Signaling field is 4 bits in length.

CRC is 16 bits.

The FHSS preamble includes an 80-bit synchronization pattern used to detect the presence of a signal, resolve antenna diversity, and acquire symbol timing, and a 16-bit Start of Frame Delimiter (SFD) that provides symbol-level frame synchronization. Concerning the latter, the SFD contains four distinct 4-bit patterns that enable the results of all possible GFSK modulations to be verified.

The FHSS frame header consists of three fields. The Length field, which is a 12-bit field, indicates the length of the payload field. Thus, the maximum length of the variable data field is 4095 bytes.

The second field in the header is a 4-bit Signaling field. One bit in this field indicates whether the data rate is 1 or 2 Mbps, while the other three bits are reserved for future use. Finally, the third field in the header is a 16-bit *Cyclic Redundancy Check* (CRC), which uses the ITU-T generating polynomial shown below:

$$G(X) = X^{16} + x^{12} + x^5 + 1$$

This polynomial is a shorthand representation for the bit sequence 1000100000010001.

Although FHSS can operate at either 1 or 2 Mbps, the preamble and header are always transmitted at 1 Mbps. Then, if the Signaling field indicates 2 Mbps operation, the remainder of the frame is transmitted at that operating rate.

Hopping Sequence

To minimize the effect of multipath reflections, the frequency used by FHSS was developed to provide a minimum hop distance. With a minimum frequency variation between hops, reflections from one hop have minimal effect on the next hop because it takes time for reflections to arrive at a receiver, which should be "looking" to receive information at a different frequency.

In the United States and Canada, a predesignated sequence of seventy-nine hop frequencies is used by FHSS, with a minimum hop distance of six channels. The base frequency is 2402 MHz and a pseudorandom number in the range zero to seventy-eight is used to select the next frequency by adding that frequency to the previously used frequency and performing a mod-79 operation on the result. For example, we can represent FHSS frequency used as 2402 + b(I), where b(I) is the base sequence in the range zero to seventy-eight. Then, the *j-th* sequence is formed from the base sequence as follows:

$$2402 + (b(i) + j) \bmod 79$$

Direct Sequence Spread Spectrum

Continuing our examination of the IEEE 802.11 specification, a second physical layer supported by the standard is direct sequence spread spectrum (DSSS). In DSSS, a signal is spread through the use of a pseudorandom sequence, which results in the use of a wider bandwidth with a lower power density.

Overview

The IEEE 802.11 DSSS standard operates at either 1 or 2 Mbps in the 2.4 GHz ISM frequency band. At the physical layer, each information bit is modulated by an 11-bit sequence referred to as an 11-bit *Barker sequence*. The use of the Barker sequence results in an 11-MHz chipping rate and spreads RF energy across a wider bandwidth than would be required to transmit the original information bits. At the receiver, the 11-bit Barker sequence is used to despread the RF input, thus enabling the original data to be recovered.

Modulation

Two different modulation methods are supported by the 802.11 DSSS standard. At an operating rate of 1 Mbps *differential binary phase shift keying* (DBPSK) is used. At an operating rate of 2 Mbps *differential quadrature phase shift keying* (DQPSK) is employed.

Using DBPSK modulation, each bit is represented by one of two possible phase changes. In comparison, in DQPSK the modulation process operates on pairs of bits, modulating each bit pair into one of four possible phase changes.

Frequency Allocation

Much as in FHSS, the frequency allocations for DSSS can vary based on different regulatory agencies. As an example of potential differences, Table 6.2 lists the allowed center frequencies and the corresponding channel numbers for North America, Europe, and Japan, which represent the three major markets for the use of DSSS wireless LANs.

TABLE 6.2

Examples of Direct Sequence Spread Spectrum Frequency Utilization

Channel Number	Frequencies in MHz		
	North America	Europe	Japan
1	2412	N/A	N/A
2	2417	N/A	N/A
3	2422	2422	N/A
4	2427	2427	N/A
5	2432	2432	N/A
6	2437	2437	N/A
7	2442	2442	N/A
8	2447	2447	N/A
9	2452	2452	N/A
10	2457	2457	N/A
11	2462	2462	N/A
12	N/A	N/A	2484

Frame Format

Figure 6.7 illustrates the general DSSS frame format. Note that the preamble and header are always at a data rate of 1 Mbps modulated through the use of DBPSK. Much like the operation of FHSS, the key to the ability to indicate the operating rate of DSS is obtained from the Signal field. This field was initially used to indicate operating rates of 1 or 2 Mbps under the 802.11 standard. With the introduction of the 802.11b standard the revision to the Signal field allowed support for operating rates of 5.5 and 11 Mbps, which are additions specified by the new specifications.

Figure 6.7
DSSS frame format.

Bits	Preamble (128)	SFD (16)	Signal (8)	Service (8)	Length (16)	CRC (16)	Data

The DSSS Preamble field consists of 128 bits and provides a mechanism for the receiving station to adjust to the incoming signal. This field is followed by the Start of Frame Delimiter (SFD) field. This 16-bit field is followed by an 8-bit Signal field. As previously mentioned, the Signal field functions as a rate indication method that allows the receiver to use the applicable modulation method commensurate with the data rate of the originator.

The fourth field in the DSS physical layer is the Service field. This 8-bit field is currently assigned the value hex 00 to signify 802.11 compliance; however, its actual use is presently reserved.

The fifth field is the Length field. The function of this 16-bit field is to indicate the number of bytes, in the Data field, that follow the CRC field. The CRC 16-bit field is used to protect the Signal, Service, and Length fields.

Infrared

In concluding our examination of the IEEE 802.11 physical layer, we examine the third physical layer supported by the 802.11 specification, the infrared physical layer.

The infrared transmission employed under the 802.11 specification is based on the 850 nm to 950 nm range, which is nearly visible light. Infrared reception is based on diffused infrared transmission, which means that a clear line-of-sight path between transmitter and receiver is not required. The allowable range between stations is limited to approximately 10 meters, however, and the use of this layer is restricted to in-building applications.

Modulation

Under the IEEE 802.11 standard, infrared communications occurs using two types of *pulse position modulation* (PPM). Under PPM the symbol period is broken into n time slots or intervals referred to as *chips*. A narrow pulse of infrared radiation at a wavelength between 850 and 950 nm is then transmitted in one of the time slots.

There are two PPM methods used under the IEEE 802.11 standard. At a data rate of 1 Mbps, sixteen-position PPM (16-PPM) is employed. Using this version of PPM, four data bits are mapped into four time slots within a sequence of sixteen time slots.

Figure 6.8 illustrates an example of 16-PPM under which four bits are mapped into applicable time slots to form a symbol. In the example illustrated in Figure 6.8 the quad-bits mapped result in a value of 9. The use of PPM is based on the fact that it provides a power conservation method in comparison to the use of *on-off keying* (OOK). In OOK a pulse is transmitted if a code bit is 1 while a 0 is represented by the absence of a pulse. In PPM, a constant is transmitted within only one of the chips while the remaining chips have no power.

A second version of PPM used for infrared communications at 2 Mbps is 4-PPM. 4-PPM maps two data bits into one of four pulses. An example of 4-PPM is illustrated in Figure 6.9. Much like 16-PPM, 4-PPM has a 250 ns slot time. When 16-PPM is employed, four bits can be transmitted in 16 slots $\times$ 250 ns/slot or 4 ms, which results in a data transmission rate of 1Mbps. In comparison, the use of 4-PPM results in eight data bits being transmitted during a period of 4 ms, resulting in a data transmission rate of 2 Mbps.

Figure 6.8
At 1 Mbps infrared communications uses 16-PPM modulation.

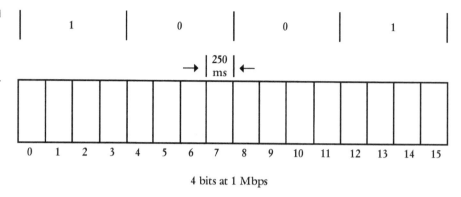

4 bits at 1 Mbps

Figure 6.9
4-PPM results in a data transmission rate of 2 Mbps.

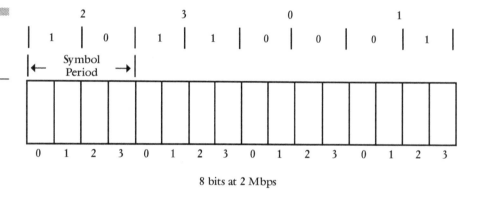

8 bits at 2 Mbps

Frame Format

Figure 6.10 illustrates the infrared physical layer frame format. The Synchronization and Start of Frame Delimited (SFD) fields function similarly to those fields used in the FHSS and DSSS frames. However, instead of a Signaling field, the Infrared frame uses a Data Rate field to denote the data rate. This field is used to indicate if the operating rate is the 1 Mbps basic access rate or the enhanced access rate of 2 Mbps.

If you compare the frame formats for FSSS and DSSS frames to the infrared frame format, you will note that the Signal field in the first two frames is in the same location as the Data Rate frame in the infrared frame. Thus, although the terminology changed, the fields are positioned and function in the same manner.

Figure 6.10
Infrared layer frame format.

Preamble	SFD	Data Rate	DC Level Adjustment	Data

The MAC Layer

In this section we literally rise up the protocol stack as we turn our attention to the Media Access Control (MAC) layer. The MAC layer represents a uniform scheme that supports the multiple physical layers indicated in Figure 6.1. Although the primary function of the MAC layer is to control access to the wireless environment, this layer is also responsible for other functions. Those additional functions can include fragmentation, encryption, power management, and synchronization. In addition, the MAC layer is responsible for providing roaming support where there are multiple access points.

Basic Access Method

The IEEE 802.11 standard uses a variation of the *Carrier Sense Multiple Access with Collision Avoidance* (CSMA/CA) protocol to provide a wireless access capability. The CSMA/CA protocol avoids collisions, instead of detecting a collision as the CSMA/CD protocol used by the IEEE 802.3 (Ethernet) standard does.

The variation of the CSMA/CA protocol used requires a station that has information to transmit first to "listen" to the medium. If the medium is busy, the station defers its transmission. If the medium is available for a specified time, referred to as the *Distributed Inter Frame Space* (DIFS), the station can transmit. Because it is possible that another station could transmit at approximately the same time, the receiver checks the CRC of received packets and transmits an acknowledgment that serves as an indicator to the originator that no collision occurred. Otherwise, if the sender does not receive an acknowledgment, it retransmits until it either receives an acknowledgment or a predefined number of retransmissions occur. Concerning the latter, if the sender cannot receive an acknowledgment after a fixed number of tries, it abandons its effort and the higher layer in the protocol stack governs how the inability to transfer data is handled.

The access method used by the IEEE 802.11 standard is officially referred to as the *Distributed Coordination Function* (DCF), which is considered to represent a CSMA/CA protocol. The reason for the selection of an access scheme with an acknowledgment instead of the near-ubiquitous wired LANs Ethernet *Carrier Sense Multiple Access/Collision Detection* (CSMA/CD) scheme is that the latter is impractical for a wireless environment. A collision detection method requires a full-duplex radio frequency or infrared pair of channels, which would be costly. In addition, unlike in a wired LAN, where it is assumed that all stations can hear a collision, in a wireless environment this is not always true. Thus, the IEEE had a sound basis for bypassing CSMA/CD for a CSMA/CA scheme, which incorporates a positive acknowledgment.

Minimizing Collisions

Because it is possible that two stations can both listen and, hearing no activity, transmit, or two stations do not hear one another and both transmit, collisions can occur. To reduce the probability of collisions, the CSMA/CA derivative used by the IEEE 802.11 standard employs a technique referred to as *virtual carrier sense* (VCS). Under VCS, a station that needs to transmit information first transmits a *Request to Send* (RTS) packet. The RTS packet represents a relatively short control packet that contains the source and destination address and the duration of the following transmission. The duration is specified in terms of the time for the transmission of the data packet and the acknowledgment of the packet by the receiver. The receiver responds to the RTS packet with a *Clear to Send* (CTS) control packet. The CTS packet indicates the same duration information as contained in the RTS control packet. Each station that receives either the RTS, CTS control packet, or both, sets its virtual carrier sense indicator for the duration of the transmission. The VCS indicator is called the *network allocation vector* (NAV) under IEEE 802.11 and serves as a mechanism to alert all other stations on the medium to back off or defer their transmission.

If the CTS is not received within a predefined period, the originating station assumes a collision occurred and initiates the procedure again, issuing another RTS packet. Once the CTS frame is received and a data frame is sent, the receiver returns an ACK packet to acknowledge a successful data transmission.

The use of the RTS and CTS control packets reduces the probability of a collision occurring at the receiver from a station "hidden" from the transmitter. Here the term "hidden node" refers to a station in a service set that cannot detect the transmission of another station and thus fails to recognize that the medium is busy. To illustrate how this can happen, consider Figure 6.11, which illustrates transmission among three stations. In this example, stations A and B can communicate; however, an obstruction prevents station C from receiving the transmission of station A. Thus, station C would not know the channel is busy and A and B could attempt to transmit at the same time. To resolve the hidden node problem, the station hearing the CTS control packet will "reserve" the medium as busy until the transaction is completed. The duration information in the RTS control packet also serves to protect the transmitter area from collisions during the ACK. In addition, because the RTS and CTS control packets are relatively short, their use reduces the overhead of collisions, because they are recognized more quickly than if the collision occurred when a normal data packet was transmitted.

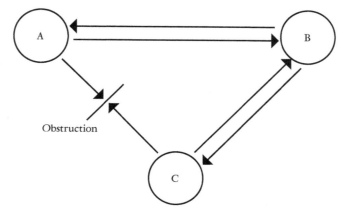

Figure 6.11
An example of the wireless "hidden" node transmission problem.

An obstruction prevents C from hearing A's transmission.
Thus, C cannot determine the channel is busy, which could
result in A and C attempting to transmit to B at the same time.

We can visually summarize the use of RTS and CTS control packets and their relationship to the flow of data and the NAV, as illustrated in Figure 6.12. In examining Figure 6.12, note that the duration fields in the RTS and CTS frames function as a mechanism to distribute medium reservation information contained as duration data in a field in each frame. That duration data is stored as a network allocation vector and indicates the period when the medium is busy.

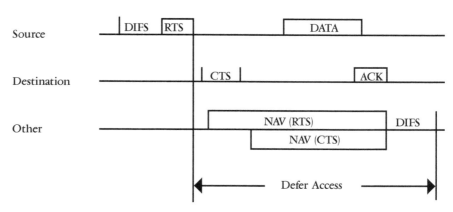

Figure 6.12
The relationship between RTS, CTS, ACK and Data packets.

Legend:
NAV Network Allocation Vector

Interframe Spaces

The DIFS represents a distributed interframe space and is one of four time gaps or interframe spaces defined to separate transmissions. The DIFS is the interframe space used for a station willing to begin a new transmission. The other three interframe spaces include the *Short Interframe Space* (SIFS), *Point Coordination IFS* (PISF), and *Extended IFS* (EIFS).

The SIFS is employed to separate transmissions belonging to an immediate action, such as the transmission of an ACK, RTS, or CTS frame. The SIFS is the shortest IFS and its value depends on the physical layer employed. For example, DSSS uses a 10 ms SIFS; its value is 28 and 7 ms when FHSS and IR are used.

The PIFS is used by an access point to gain access to the medium. The value of the PIFS is computed as the PIFS plus one slot time. Similar to SIFS, the PIFS interval varies based on the physical layer. Its duration is 30 ms for DSSS, 78 ms for FHSS, and 23 ms for IR.

The DIFS is computed as the PIFS plus one time slot. For DSS the DIFS is 50 ms; its value is 128 ms for FHSS and 23 ms for IR.

The fourth type of Interframe Space is the Extended IFS (EIFS). The EIFS represents a delay by a station that received a frame it could not understand. The EIFS prevents a station that could not understand the duration of the virtual carrier sense (VCS) from colliding with a subsequent frame belonging to the current dialog. Table 6.3 summarizes the interframe delays and slot times for the three physical layers supported by the 802.11 specification.

TABLE 6.3

Interframe Spaces

	Physical Layer		
Interframe Space	**DSSS**	**FHSS**	**IR**
Slot time	20 μs	50 μs	8 μs
SIFS	10 μs	28 μs	7 μs
PIFS	30 μs	78 μs	15 μs
DIFS	50 μs	128 μs	23 μs

Collision Avoidance

A key to the avoidance of collisions is the use of an exponential *backoff algorithm*. When a station detects that the medium is busy, it first delays activity until the end of the DIFS interval. At this point, the station waits a given number of time slots based on the exponential backoff algorithm. Using this algorithm, each station selects a random number and waits that number of slots prior to accessing the medium; however, the station will obviously check to determine if a different station has access to the medium prior to accessing that medium. The "exponential" portion of the algorithm means that each time a station selects a slot number and has a collision, it increases exponentially the maximum number for the random selection process.

There are three conditions when the exponential backoff algorithm must be executed. First, a station that senses the medium is busy prior to transmitting a frame will execute the algorithm. The algorithm will also be used after each successful transmission and after each retransmission. Only when a station requires the ability to transmit a new frame, and the medium has been available longer than the DIFS, does a station avoid using the algorithm.

Now that we have an appreciation for the manner in which the MAC layer controls access to the medium, let us turn our attention to the types of layer 2 frames and their formats.

Frame Types

The MAC layer supports three main types of frames—data frames that are used to transmit information between stations, control frames that are used to control access to the medium, and management frames. The last are used to exchange management information between stations at layer 2 but are not forwarded to upper layers in the protocol suite.

FRAME FORMATS. The format of 802.11 frames varies. Figure 6.13 illustrates the format of the MAC data frame used to transmit information between stations. As we note later in this section, several fields from this frame are used in other types of frames.

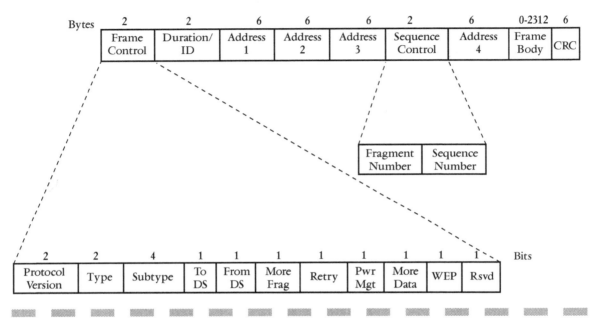

Figure 6.13 The general MAC data frame format.

In examining Figure 6.13, note that the Frame Body field can be a maximum of 2312 bytes in length. This provides the ability to transport a maximum length Ethernet frame, which has a 1500-byte Information field. Because the bit error rate of a radio link is much higher than that of a LAN, however, the probability of a frame's being corrupted is higher as the length of a frame increases. This is much more the case on a wireless LAN than on a wired LAN. To compensate for this situation, a simple fragmentation and reassembly mechanism is supported at the MAC layer.

CONTROL FIELD. The 16-bit Frame control field consists of eleven fields; of these, eight are 1-bit fields are set to indicate a specific feature or function is enabled. In this section we examine the use of each subfield within the Control field.

PROTOCOL VERSION SUBFIELD. The 2-bit Protocol Version subfield provides a mechanism to identify the version of the 802.11 standard. In the initial version of the standard the Protocol Version field value is set to 0.

TYPE AND SUBTYPE SUBFIELDS. The Type and Subtype subfields provide 6 bits that identify the frame. The Type subfield is capable of identifying four types of frames; however, only three types are presently defined. The 4-bit Subtype subfield identifies a specific type of frame within the Type category.

Table 6.4 lists the Type and Subtype subfield values and a description of what the values of the 6-bit positions indicate. In examining the entries in the Subtype Description column, note that the term Beacon has nothing to do with a Token-Ring network. Instead, a beacon frame is periodically transmitted by an access point with the value of its clock at the time of transmission. This allows receiving stations to stay in synchronization with the AP's clock.

TODS SUBFIELD. This 1-bit field is set to a value of 1 when the frame is addressed to an AP for forwarding to the Distribution System. Otherwise, the bit is set to a value of 0.

FROMDS SUBFIELD. The FromDS is also a 1-bit subfield. The value of this field is set to 1 when the frame is received from the Distribution System. Otherwise, the field value is set to 0.

MORE FRAGMENTS SUBFIELD. This subfield is another 1-bit field. The value of this field is set to 1 when there are more fragments following the current fragment. Thus, this field lets the originator note that a frame represents a fragment and allows the receiver to reconstruct a series of fragments into a frame.

TABLE 6.4

Type and Subtype
Subfield Values

Type Value	Type Description	Subtype Value	Subtype Description
b3 b2		b7 b6 b5 b4	
00	Management	0001	Association Request
00	Management	0001	Association Response
00	Management	0010	Association Request
00	Management	0011	Reassociation Response
00	Management	0100	Probe Request
00	Management	0101	Probe Response
00	Management	0110-0111	Reserved
00	Management	1000	Beacon
00	Management	1001	ATIM
00	Management	1010	Disassociation
00	Management	1011	Authentication
00	Management	1100	Deauthentication
00	Management	1101-1111	Reserved
01	Control	0000-0001	Reserved
01	Control	1010	PS-Poll
01	Control	1011	RTS
01	Control	1100	CTS
01	Control	1101	ACK
01	Control	1110	CF End
01	Control	1111	CF End + CF-ACK
10	Data	0000	Data
10	Data	0001	Data + CF-ACK
10	Data	0010	Data + CF-Poll
10	Data	0011	Data + CF-ACK + CF-Poll
10	Data	0100	Null Function (no data)
10	Data	0101	CF-ACK (no data)
10	Data	0110	CF-Poll (no data)
10	Data	0111	CF-ACK + CF-Poll (no data)
10	Data	1000-1111	Reserved
10	Data	0000-1111	Reserved

Figure 6.14 illustrates the frame fragmentation process. Note that fragments 0, 1, and 2 have their More Fragments subfield values set to 1 in the MAC header in each fragment in this example.

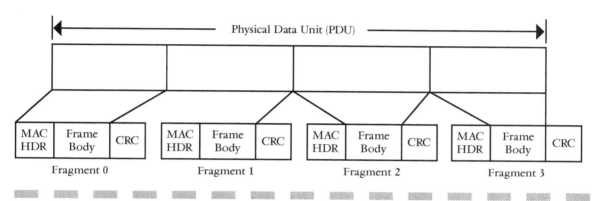

Figure 6.14   *Frame fragmentation.*

The fragmentation transmission process under the IEEE 802.11 standard is based on a simple *send-and-wait algorithm*. Under this algorithm the transmitting station cannot send a new fragment until it either receives an ACK for the prior segment or decides that the fragment was retransmitted a predefined number of times and drops the entire frame.

RETRY SUBFIELD.   This 1-bit field is set to indicate that the frame is a fragment representing the retransmission of a previously transmitted fragment. The receiving station uses this field to recognize duplicate transmissions that can occur if an Acknowledgment packet is lost.

POWER MANAGEMENT SUBFIELD.   IEEE 802.11 stations can be in one of two power modes—*Power Save* or *Active*. A station that is Active when transmitting a frame can change its power status from Active to Power Save.

Through the use of the Power Management bit, a station can indicate its power state. This information is used by the Access

Point, which continuously maintains a record of stations working in the Power Saving mode. The AP then buffers packets addressed to those stations until they either specifically request packets by transmitting a polling request or they change their power state.

Another technique used to transmit buffered frames to a station in its Power Save mode of operation is obtained through the use of Beacon frames. The AP periodically transmits information concerning which stations operating in a Power Saving mode have frames buffered by the Access Point as part of its Beacon frames. Each such stations then wakes up to receive the Beacon frame and notes that there is a frame stored at the AP awaiting delivery. The station then remains in an Active power state and transmits a polling message to the AP to retrieve those frames.

MORE DATA SUBFIELD. As its name implies, the More Data subfield indicates there are more frames following the current frame. This 1-bit subfield is set by the AP to indicate there are more frames buffered to a particular station. Remember that buffering at the AP occurs when a destination station is in its Power Save mode of operation. The destination station can then use this information to decide if it should continue polling or if the station should change its power management state to Active.

WEP SUBFIELD. The IEEE 802.11 committee responsible for the development of the wireless standard addressed security through the addition of authentication and encryption collectively referred to as *Wired Equivalent Privacy* (WEP). The setting of the WEP subfield indicates that the body of the frame is encrypted according to the WEP algorithm.

WEP uses the RC4 encryption algorithm, which is a stream cipher. For those not familiar with encryption, a *stream cipher* operates by expanding a short key into an infinite pseudorandom key stream. The transmitter XORs (modulo-2 adds) the key stream to the plaintext, resulting in the generation of encrypted ciphertext. As you might expect, the receiver must have a copy of the key to be able to correctly convert the ciphertext back to plaintext. In doing so, the receiver uses the key to generate the same sequence of pseudorandom bits; these are then modulo-2 subtracted from the ciphertext to reconstruct the plaintext. Figure 6.15 illustrates a simple example of encryption and description based on modulo-2 operations.

Figure 6.15
An example of
encryption and
decryption based on
modulo-2 arithmetic.

plaintext	1010110	ciphertext	11000110
pseudo-random bits	01101011	pseudo-random bits	01101011
Modulo-2 addition	11000110	Modulo-2 subtraction	10101101
Forms ciphertext		Restores plaintext	

The WEP algorithm uses a pseudorandom number generator that is initialized by a 40-bit secret key. Through the use of a 40-bit security key and a 24-bit initialization vector, a 64-bit key is generated. Most products, such as access points and network adapter cards, include a WEB configuration screen that prompts the user to type ten hexadecimal digits (any combination of 0—9, a—f, or A—F) for the 40-bit WEP. Some equipment supports 128-bit WEP keys, requiring the entry of twenty-six hexadecimal digits. Both the access point and clients supported by the access point must be configured to use the same encryption key and key length.

Once you enter an applicable sequence of digits, those digits are then used to generate a key sequence of pseudorandom bits whose length is equal to the largest possible packet. Those bits are then modulo-2 added to the frame bits to encrypt the frame. Each frame is transmitted with an initialization vector, which restarts the pseudorandom number generator to provide a new key sequence for the subsequent frame. Thus, this technique is difficult for a brute force attack to compromise. Because a station must have knowledge of the key to correctly decrypt data, the key in effect becomes an authentication mechanism.

There are several problems associated with WEP that came to light during 2001 based on a research effort mounted by the University of California–Berkley. Because WEP is based on a stream cipher, it is possible for an eavesdropper to intercept two cipher-texts encrypted with the same key stream and obtain the XOR of the two plaintexts. A knowledgeable person can use statistical analysis to recover the plaintext. Another problem with WEP is that it is possible to change the value of information in such a way that a correct CRC is generated for the modified data frame. In spite of these problems, perhaps the key problem is the fact that WEP, by default, is disabled. This means that an organization that simply installs one or more access points and network cards in client computers without enabling WEP is, in effect, naked to prying eyes. Although an important paper was published during 2001 outlining WEP security vulnerabilities, it is a lot easier to denote potential problems than to actually break into a system. Thus, although WEP certainly has some weaknesses, it is impor-tant to note that you can overcome many of those weaknesses. First, you can consider using a 128-bit key. In doing so, however, you must note that the key length is not standardized. The use of a 128-bit key will more than likely result in your organization's being restricted to a single vendor. A second method to facilitate security is to employ access points that add authentication. For example, Agere Systems offers an access point that can store up to 492 MAC addresses having permission to access the network. If a client with a MAC address that was not added to the access point attempts to access the AP it will be rejected, in effect blocking the client from gaining access to the network. This technique, although effective for blocking unauthorized users, still leaves WEP encryption vulnerabilities unaffected. In addition, if a per-son is able to read on-the-air traffic, that person can note source addresses and, with a bit of effort, overcome the barrier. From a practical standpoint you must examine where your access points and mobile users are located and identify potential obstructions that could make it difficult, if not impossible, for unwanted peo-ple to sit in a van in a parking lot and "read" the activity on your network. Another technique you can consider to limit your orga-nization's vulnerability is to note that an access point in effect is a bridge that broadcasts LAN traffic onto the air. Recognizing this, you can place an access point on a nonpopulated hub and con-nect the hub to a backbone router. This network configuration limits hub traffic to frames explicitly destined to mobile clients, in effect hiding interwired LAN traffic from the air. Thus, proper planning can overcome many of the problems associated with transmitting data in a wireless environment as well as the use of WEP.

ORDER SUBFIELD.   The last position in the Control field is the 1-bit Order subfield. The setting of this bit indicates that the frame is being transmitted using the *Strictly Ordered* service class. The use of this bit position accommodates the DEC LAT protocol that cannot accept change of ordering between unicast and multicast frames. Thus, for the vast majority of wireless applications this subfield is not used.

Now that we have an appreciation for the subfields within the control field, let us continue our examination of the MAC Data frame.

DURATION/ID FIELD.   The meaning of this field depends on the type of frame. In a Power-Save Poll message, this field indicates the station identification (ID). In all other types of frames, this field indicates the Duration value, which represents the time in microseconds requested to transmit a frame and its interval to the next frame.

ADDRESS FIELDS.   A frame can contain up to four addresses, depending on the setting of the ToDS and FromDS bits in the Control field. Those address fields are labeled Address 1 through Address 4.

The use of the address fields, based on the settings of the ToDS and FromDS bits in the Control field, are summarized in Table 6.5. In examining Table 6.5, note that Address 1 always indicates the recipient address. That address can be the destination address (DA), the Basic Service Set ID (BSSID), or the recipient address (RA). If the ToDS bit is set, Address 1 contains the AP address. When the ToDS bit is not set, the value of the Address 1 field contains the station address. All stations filter on the Address 1 field value.

TABLE 6.5

MAC Data Address
Field Values Are
Based on ToDS
and FromDS Bit
Settings in the
Control Field

ToDS	FromDS	Address 1	Address 2	Address 3	Address 4
0	0	DA	SA	BSSID	N/A
0	1	DA	BSSID	SA	N/A
1	0	BSSID	SA	DA	N/A
1	1	RA	TA	DA	SA

TA = Transmitter Address
RA = Receiver Address
BSSID = Basic Service Set ID

Address 2 is always used to identify the station transmitting the packet. If the FromDS bit is set, the value in the Address 2 field is the AP address; otherwise it represents the station address. The Address 3 field also depends on the ToDS and FromDS bit settings. When the FromDS bit in the Control field is set to a value of 1, the Address field contains the original Source Address (SA). If the frame has the ToDS bit set, then the Address 3 field contains the Destination Address (DA).

The Address 4 field is used for the special situation where a wireless distribution system is employed and a frame is being transmitted from one access point to another. In this situation both the ToDS and FromDS bits are set. Thus, both the original Destination and the original Source addresses are not applicable and Address 4 is limited to identifying the source of the wired DS frame.

SEQUENCE CONTROL FIELD.   The 2-byte Sequence Control field functions as a mechanism to represent the order of different fragments that are part of the frame. As previously illustrated in Figure 6.13, the Sequence Control field consists of two subfields— Fragment Number and Sequence Number. Those subfields are used to define the frame and the number of the fragment that is part of a frame.

FRAME BODY FIELD.   The Frame Body field is used to transport actual information between stations. As indicated in Figure 6.13, this field can vary in length up to 2312 bytes.

CRC FIELD. The last field in the MAC Data frame is the CRC field. This field is 4 bytes in length and is used to contain a 32-bit CRC. Now that we understand the format of the MAC Data Frame, let us turn our attention to the composition of other common frames.

RTS Frame

Figure 6.16 illustrates the RTS frame format. Note that the MAC header consists of the first four fields of that frame. The Recipient Address in the frame represents the address of the wireless network station that is the intended immediate recipient of the next data or management frame. The transmitter address (TA) represents the address of the station transmitting the RTS frame, whereas the Duration field contains the time in microseconds (μs) required to transmit the next data or management frame plus one CTS frame, one ACK frame, and three interval periods between frames. Also note that the RA and TA fields are precisely the same length as the Source and Destination fields included in IEEE 802.3 Ethernet wired LAN frames.

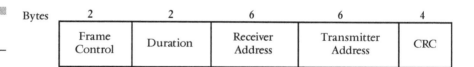

Figure 6.16
The RTS frame.

Bytes	2	2	6	6	4
	Frame Control	Duration	Receiver Address	Transmitter Address	CRC

CTS Frame

Because the CTS frame is issued in response to the receipt of an RTS frame, there is a relationship between certain fields in each frame. To observe this relationship, consider Figure 6.17, which illustrates the CTS frame format.

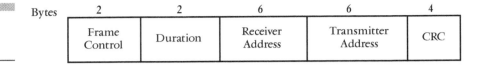

Bytes	2	2	6	6	4
	Frame Control	Duration	Receiver Address	Transmitter Address	CRC

In examining Figure 6.17 and recognizing that the CTS frame responds to an RTS frame, it comes as no surprise that the Receiver Address (RA) of the CTS frame is copied from the Transmitter Address (TA) field of the received RTS frame. The duration field value is obtained from the Duration field of the previously received RTS frame less the time, in microseconds, required to transmit the CTS frame and the Short Interframe Space (SIFS) interval. The RA and TA fields in the CTS frame are both 48 bits in length and represent the address length used by IEEE 802.3 wired LANs.

ACK Frame

The third commonly used short frame is the ACK frame. The format of this frame is shown in Figure 6.18.

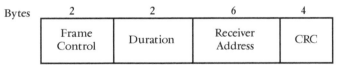

Bytes	2	2	6	4
	Frame Control	Duration	Receiver Address	CRC

As in the CTS frame, several fields in the ACK frame have values based on previously received frames. For example, the Receiver Address of the ACK frame is copied from the Address 2 field of the previously received frame that the ACK acknowledges. Another example of field relationship between frames concerns the setting of the More Fragment bit in the Frame Control field of the previous frame. If that bit was set to 0, the Duration field in the ACK frame is set to 0. Otherwise, the duration field value is obtained from the Duration field of the previous frame, minus the time in microseconds required to transmit the ACK frame and its SIFS interval.

Operation

In concluding our examination of the IEEE 802.11 standard we examine two key wireless LAN operations, the method by which a station joins a basic service set and how roaming is accomplished. The first involves a station's joining an existing cell, while the second involves the process of moving from one cell to another.

Joining an Existing Cell

There are three periods when a station needs to access an existing basic service set (BSS). Those periods include the time a station is powered up, after it enters sleep mode, or when it enters a BSS area. For each situation the station must obtain synchronization information. Normally this information is obtained from an AP; however, it can also be obtained from other stations within the BSS. The latter is referred to as an *ad-hoc mode* of operation. In either situation, the station acquires synchronization by either active or passive scanning. Once this is accomplished the station must then go through an authentication process.

ACTIVE SCANNING. Active scanning requires a station to attempt to locate an access point so that it can receive synchronization information from that device. To accomplish this task, the station transmits Probe Request frames and waits for a Probe Response packet transmitted by an AP.

PASSIVE SCANNING. A station can obtain synchronization information via passive scanning. Assuming a station is within an applicable distance of an AP, it can listen for a Beacon frame that is periodically transmitted by each AP. Because the Beacon frame contains synchronization information, the ability to listen to Beacon frames provides a station with the synchronization information it needs. The actual method used by a station, either active or passive scanning, primarily depends on two factors, the power consumption of the station and its performance.

Authentication and Association

Once a station locates an access point and obtains synchronization information, it must exchange authentication information. This interchange of information occurs between the AP and the station, with each device providing knowledge of a given password.

After the station is authenticated the association process commences. Under the association process, information about the stations and the capabilities of the BSS is exchanged. This allows a group of APs to obtain information about the current position of a station. Once the association process is completed, the station is capable of transmitting and receiving frames. Figure 6.19 illustrates the sequence of steps in the active scanning and association process required prior to a station's being able to exchange information. Note that although two access points are shown, a station could examine more than two Probe Response frames to select an AP.

Now that we have an appreciation for the manner by which a station can join a cell, let us investigate second key operation, roaming.

Roaming

Each station on an 802.11 wireless LAN is associated with a particular access point. When a station moves from one cell (BSS) to another without losing a connection, it is in the process of *roaming*.

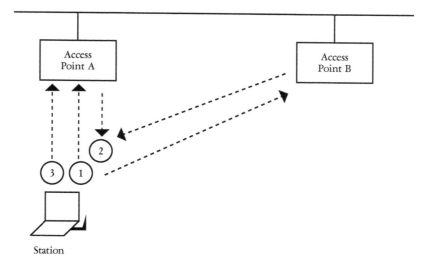

Access
Point A

Access
Point B

Station

Figure 6.4 illustrated an example of roaming. In that example
the wireless LAN infrastructure consists of three access points
and eight stations. Stations 1, 2, and 3 are associated with access
point B, while stations 7 and 8 are associated with access point C.
In that example, station 3 is shown in the process of roaming by
moving from the area of coverage of access point A to the area of
coverage of access point C.

COMPARISON TO CELLULAR.   Although roaming in a wireless LAN environment is similar to roaming in a cellular environment, there are two key differences. First, the 802.11 standard supports the transmission of packets that have a destination address,
sequence, and fragment identification. This facilitates LAN roaming because the transition can be expected to occur at a "walking"
rate instead of in a fast-moving vehicle. Second, although a temporary gap in transmission may not affect a voice conversation, its
effect can be more pronounced in a wireless LAN environment
because the inability to receive a packet causes the originating station to perform a retransmission. This is typically initiated by an
upper-layer protocol that sets a timer prior to each transmission.
If the timer expires without the transmitting station receiving an
acknowledgment, it retransmits the packet. Because real-time

voice is simply lost and is not retransmitted, a temporary disconnection can adversely affect a wireless LAN more than it would a cellular phone network.

OPERATION. When roaming occurs, a station moving away from its access point notes that its link to the AP is becoming poor. The station then uses its scanning function to locate another AP or it could use information from a previous scan to select another AP. Once a new AP is located, the station transmits a Reassociation Request to the AP. If the station receives a Reassociation Response, the station has a new AP and has successfully roamed.

To generate a Reassociation Response the AP must indicate the reassociation to the Distribution System (DS). The DS updates its information and notifies the old AP that it is no longer associated with the station. If the station does not receive a Reassociation Response, it scans for another AP. Although this provides the general approach to roaming, the actual implementation method is left to the product developer.

Now that we have an appreciation for the utilization and operation of the IEEE 802.11 wireless LAN, we conclude this chapter by turning our attention to the two extensions to this wireless standard.

The 802.11b Standard Extension

The second wireless LAN standard defined by the IEEE is the first extension to the 802.11 standard. This extension is referred to as the 802.11b standard. In comparison to the original 802.11 standard, which defined three physical layers, the 802.11b standard is restricted to the support of DSSS. In this section we focus our attention on the 802.11b standard and its support of DSSS. This takes us to new heights, because the data transfer rate supported by DSSS increases to 11 Mbps.

Overview

The IEEE 802.11b standard dates to September 1999, when the 802 committee extended the original 802.11 specification. The resulting extension, which is referred to as 802.11b, is restricted to direct sequence spread spectrum (DSSS). Unlike the 802.11 standard, where DSSS is limited to data rates of 1 and 2 Mbps, the 802.11b standard adds data transmission rates of 5.5 Mbps and 11 Mbps through the use of a new modulation method.

Operation

Much like the 802.11 standard, the 802.11b standard supports an equivalent set of features including operation in the 2.4-GHz band frame format, the use of WEP (with a default of disabled), and implementation in products including access points and PC cards. Where the 802.11b standard differs from the 802.11 standard is in the manner by which modulation occurs and the available transmission distance in an indoor environment. The 802.11 standard uses an 11-bit Barker chip as a spreading mechanism whereas the 802.11b specification is based on the use of an 8-bit *complementary code keying* (CCK) algorithm. Although it is theoretically possible to transmit at distances up to 500 feet in an indoor environment under the 802.11 standard at 1 Mbps, when using 802.11b at a data rate of 11 Mbps you can more than likely expect a realistic transmission distance of 100 to 125 feet in an office environment.

Modulation

Complementary code keying supports operating rates of both 5.5 and 11 Mbps through the use of an eight chip code-spreading sequence. At 11 Mbps the delay spread of the CCK code-spreading sequence is 100 ns, whereas at 5.5 Mbps the delay spread is 250 ns.

Because the 802.11b standard is backward compatible with the 802.11 DSSS standard it also supports transmission at 1 and 2 Mbps. Thus, an organization can purchase 802.11b equipment for use in an 802.11 DSSS environment until it is ready to fully upgrade to the higher operating rate.

BARKER CODE.   Figure 6.20 illustrates the use of an 11-bit Barker code used to spread a signal. Note that this code is not only a pseudorandom (PRN) code but represents a polar code of +1 and −1 values. As indicated in Figure 6.20, each data bit is expanded by the 11-bit Barker code, resulting in eleven chips being substituted for each data bit. Each binary 1 data bit results in the direct use of the applicable 11-bit Barker code, whereas each 0 data bit results in the spreaded code representing the inversion of the 11-bit Barker code. When operating at 1 and 2 Mbps each 802.11b compliant device uses the Barker code instead of CCK to spread a signal.

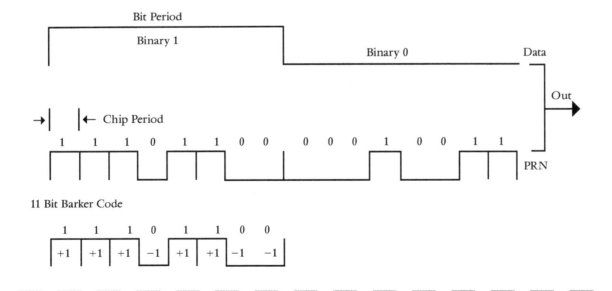

Figure 6.20   Under the IEEE 802.11 DSS specification an 11-bit pseudorandom Barker code is used to spread each data bit.

COMPLEMENTARY CODE KEYING (CCK). Complementary code keying (CCK), although representing a modern modulation method, actually dates to 1951 when the first published work on complementary sequences appeared. A paper written by Marcel Golay, described the properties of a complementary sequence used to control slits in a multislit spectrometer.

COMPLEMENTARY CODES. The basis behind CCK, as you might expect, is a complementary code, so let us first focus our attention in this area. A complementary code represents a pair of equal length sequences, such that the number of pairs of like elements with any given separation in one series is equal to the number of pairs of unlike elements with the same separation in the other series. Because the symmetry described may not be intuitive, let us examine an example of a complementary sequence. Figure 6.21 illustrates a pair of complementary sequences defined by M.J.E. Golay in his article "Complementary Series" published in the April 1961 edition of *IRE Transactions on Information Theory.* In examining Figure 6.21, note that sequence 1 has four pairs of like elements with a separation of 1. Sequence 1 also has three pairs of unlike elements with a separation of 1. Sequence 2 has four pairs of unlike elements with a separation of 1, and three pairs of like elements with a separation of 1.

From Figure 6.21 we can also compute the number of like and unlike elements for each sequence when the pair separation is increased to 2 and 3. The results of this action are summarized in Table 6.6.

TABLE 6.6

Element Pairing for Two Sequences Illustrated in Figure 6.21

	Sequence 1		Sequence 2	
Pair Separation	**Like**	**Unlike**	**Like**	**Unlike**
1	4	3	3	4
2	4	3	3	4
3	1	5	5	1

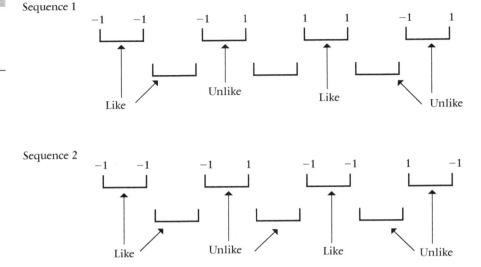

If you carefully examine the entries in Table 6.6, you cannot help but notice their symmetry. Because complementary codes have a periodic autocorrelative vector sum of zero everywhere, with the exception of at the zero shift, they become quite useful for digital communications. When applied to communications, however, the elements in the code have phase parameters and are referred to as a *polyphase complementary code*. To further make you scratch behind your ears, those complementary codes, that for communications purposes are polyphase codes, are also referred to as *spreading codes* because they are employed to spread the bandwidth used by the DSSS waveform. To eliminate confusion, I simply continue to reference CCK as is.

The IEEE 802.11b standard specifies the use of complementary spreading codes that have a code length of 8 and a chipping rate of 11 M chips/second. Through the use of a symbol rate of 1.375 M symbols/s, the 11-Mbps waveform winds up occupying approximately the same bandwidth as the 2-Mbps DSSS version that uses QPSK modulation for mapping data into symbols. Thus, one of the benefits of CCK is the fact that at a higher data rate of 11 Mbps it still permits the support of up to three nonoverlapping channels in the ISM band.

DERIVATION. The 8-bit CCK code words are derived from the following formula:

8-chip complex code $(\varnothing_1, \varnothing_2, \varnothing_3, \varnothing_4) =$

$$\left(\begin{array}{c} e^j(\varnothing_1 + \varnothing_2 + \varnothing_3 + \varnothing_4),\, e^j(\varnothing_1 + \varnothing_3 + \varnothing_4),\, e^j(\varnothing_1 + \varnothing_2 + \varnothing_4), \\ -e^j(\varnothing_1 + \varnothing_4),\, e^j(\varnothing_1 + \varnothing_2 + \varnothing_3),\, e^j(\varnothing_1 + \varnothing_3),\, -e^j(\varnothing_1 + \varnothing_2),\, e^j \varnothing_1 \end{array} \right)$$

Note that in the preceding formula the first phase $(\varnothing_1)$ is added to all code phases, the second phase $(\varnothing_2)$ is added to all odd code phases, the third phase $(\varnothing_3)$ is added to all odd pairs of code phases, and the fourth phase $(\varnothing_4)$ to all odd quads of code phases. This action represents *Hadamard encoding,* in which, for a code length of 2^n, $n + 1$ phases are encoded into 2^n output phases.

The preceding formula generates the code sets for both 11- and 5.5-Mbps data rates, with a subset 4 bits of the 11-Mbps code set used at the lower data rate. The phase parameters $\varnothing_1$ through $\varnothing_4$ represent the phase values of the complex code set and are defined in the IEEE standard.

UTILIZATION. To illustrate the use of CCK let's look at its operation at 11 Mbps. The data stream entering the encoder is used to encoded phase parameters based on dibit pairs and not the values of those pairs. This action is illustrated in Table 6.7, where $d\varnothing$ represents the least significant bit, which is the first in time to arrive at the encoder.

The actual encoding is based on differential QPSK (DQPSK) modulation. Table 6.8 illustrates the manner in which DQPSK modulation of the phase parameters occurs.

	Dibit	Phase
TABLE 6.7	(d1, d0)	0_1
CCK Phase Parameter Encoding Scheme	(d3, d2)	0_2
	(d5, d4)	0_3
	(d7, d6)	0_4

TABLE 6.7

CCK Phase Parameter Encoding Scheme

	Dibit di + 1, dj	Phase Value
TABLE 6.8	00	0
DQPSK Modulation of Phase Parameters	01	Π
	10	$\Pi/2$
	11	$\Pi/2$

TABLE 6.8

DQPSK Modulation of Phase Parameters

To illustrate an example of the generation of a CCK code word, let us assume the data stream 10110101 occurred, where the bit order is d7, d6, d∅. From Table 6.8, the pair d1, d∅, which has a value of 01, would result in $\varnothing_1$ being encoded as π. Similarly, the value of the pair d3, d2, whose value is 01, results in ∅2 = π. Continuing our coding effort, the bit pair d5, d4, whose composition is 11, results in ∅3 = +π/2, whereas the bit pair d7, d6, whose composition is 10, results in ∅4 encoded as π/2.

If we substitute the four phase parameters into the previously defined 8-chip complex code we obtain:

8-chip complex code =

$$
\left(
\begin{array}{l}
e^{j(\pi + \pi - \pi/2 + \pi/2)},\ e^{j(\pi - \pi/2 + \pi/2)},\ e^{j(\pi + \pi + \pi/2)} \\[2mm]
-e^{j(\pi + \pi/2)},\ e^{j(\pi + \pi - \pi/2)},\ e^{j(\pi - \pi/2)},\ -e^{j((\pi + \pi))},\ e^{j\pi}
\end{array}
\right)
$$

$$= \left(e^{j2\pi}, e^{j\pi}, e^{j5\pi/2}, -e^{j3\pi/2}, e^{j\pi/2}, -e^{j2\pi}, e^{j\pi} \right)$$

Because $e^{j\pi} = \cos\varnothing + j\sin\varnothing$ we obtain:

8 chip complex code =

$$\begin{pmatrix} \cos2\pi + j\sin2\pi, \cos\pi + j\sin\pi, \cos5\pi/2 + j\sin5\pi/2, \\ -\cos3\pi/2 - j\sin3\pi/2, \cos3\pi/2 + j\sin3\pi/2, \cos\pi/2 + j\sin\pi/2, \\ -\cos2\pi - j\sin2\pi, \cos\pi + j\sin\pi \end{pmatrix}$$

From the preceding we obtain the following complex code word:

$$(1, -1, j, j - j, j, -1, 1)$$

When operating at 11 Mbps, 6 bits from the 8-bit CCK code are used to select one of 64 complex codes. The remaining 2 bits are used to modulate the entire 8-chip complex code word using differential quadrature phase shift keying (DQPSK) modulation. With a symbol rate of 1.375 MHz and 8 bits encoded per symbol, this results in a data transmission rate of 11 Mbps. At 5.5 Mbps, 2 bits are used to select one of four codes, while the remaining 2 bits modulate the code, again using DQPSK. Because the symbol rate is also 1.375 MHz, when 4 bits are encoded per symbol this results in a data rate of 5.5 Mbps.

If you are puzzled about why the IEEE 802.11b standard goes to the trouble of specifying CCK, the answer is *multipath tolerance.* Because of the symmetry property of CCK, it becomes relatively easy to ignore multipath reflections. However, while multipath reflections can now be ignored, the higher data rate limits the transmission distance to a fraction of that obtainable under the basic standard.

The IEEE 802.11a Standard Extension

The second extension to the IEEE 802.11 standard is similar to the first in that it uses the same types of frames as the basic standard such as ACK, RTS, and CTS. Unlike the 802.11b standard which, like the original standard, operates in the 2.4-GHz ISM frequency band, the 802.11a standard operates in the recently allocated 5-GHz Unlicensed National Infrastructure (UNII) band. In addition to operating in a different frequency band, the 802.11a standard uses a frequency division multiplexing scheme referred to as orthogonal frequency division multiplexing (OFDM) instead of DSSS.

Overview

The use of the 5-GHz UNII frequency band provides some distinct advantages over the 2.4-GHz band. In addition to providing a greater amount of bandwidth for transmission, the 5-GHz band has less potential interference because competing technologies, such as the Bluetooth short range wireless technology, the Home RF technology, and even microwave ovens operate in the 2.4-GHz band.

FREQUENCY ALLOCATION. As a review of material previously presented in this book, the FCC allocated 300 MHz for unlicensed operations in the 5-GHz frequency band; 200 MHz is allocated from 5.15 GHz to 5.35 GHz; and the remaining 100 MHz at 5.725 GHz to 5.825 GHz. The first 100 MHz in the first portion of the first band is restricted to a maximum power output of 50 mw. The second 100 MHz is limited to a maximum power output of 250 mw. The third 100 MHz is reserved for outdoor applications and supports a maximum power output of 1w.

Modulation

The IEEE 802.11a standard uses orthogonal frequency division multiplexing. Using this modulation method, each high-speed carrier consisting of 20 MHz is broken into fifty-two subchannels, forty-eight for data and four pilot channels, each approximately 300 KHz wide. As we noted earlier when we discussed modulation methods, OFDM represents a multicarrier or discrete multitone modulation method. Under OFDM a high-speed serial data stream is segmented into a series of multiple lower-speed signals that are simultaneously transmitted in parallel over different frequencies.

The IEEE 802.11a standard defines the use of an OFDM physical layer. Data transmission rates of 6, 9, 12, 18, 24, 36, 48, and 54 Mbps are specified. At 6 and 9 Mbps, BPSK modulation is employed. At 12 and 18 Mbps, QPSK modulation is used. At 24 and 36 Mbps, 16-QAM modulation is specified, whereas at 48 and 54 Mbps 64-QAM is used.

Frame Format

Because the IEEE 802.11a standard uses a different method of modulation the frame is slightly modified. Figure 6.22 illustrates the revised IEEE 802.11a frame. This frame consists of six fields, with one field containing five subfields.

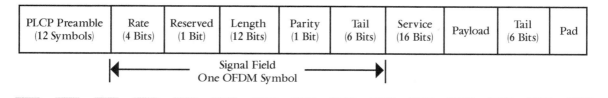

Figure 6.22   The IEEE 802.11a frame format.

PLCP FIELD. The Physical Layer Convergence Protocol (PLCP) preamble field enables the receiver to acquire an inbound OFDM signal as well as serving to synchronize the demodulator. Of the twelve symbols in the preamble field, ten are used for establishing automatic gain control and an estimate of the frequency of the carrier signal. The remaining two symbols are used by the receiver for fine tuning the signal. Through the use of the preamble the receiver can train itself to the signal 16 μs after receiving the frame.

The Signal field that follows the PLCP preamble consists of five subfields and is 24 bits in length. The length of this field represents one OFDM symbol. The Rate subfield identifies the operating rate of the frame less the PLCP and Signal fields. Both of those fields are transmitted at 6 Mbps using BPSK, regardless of the data rate indicated in the Rate field.

The Length subfield identifies the total number of bytes in the frame. This field takes a value between 1 and 4095. The 1-bit Parity subfield is based on even parity for the first 17 bits of the header, whereas the 6-bit Tail subfield has all its bit positions set to 0 to reset the convolution encoder as well as to terminate the code trellis in the decoder.

SERVICE FIELD. The Service field is 16 bits in length. The first 7 bits in this field are set to 0s to synchronize the descrambler in the receiver. The remaining 9 bits are reserved for future use and are presently set to 0.

PAYLOAD FIELD. The Payload field represents the payload from the MAC layer being transmitted.

PAD FIELD. The trailing Pad field is variable in length. This field represents the number of bits that make the payload a multiple of the number of coded bits in an OFDM symbol. As indicated earlier, when we examined modulation, the minimum number of code bits per OFDM symbol is 6; other code bits per OFDM symbol are 48, 96, 192, and 288.

Operation

The IEEE 802.11a physical layer requires a series of complex processes. Each physical data unit must be scrambled to prevent long runs of 0s and 1s in the input data being input to the remainder of the modulation process. The scrambled data is then input into a convolutional encoder and interleaved to prevent error bursts from being input to the convolutional decoding process at the receiver. Once this is accomplished, the interleaved data is mapped into an applicable symbol based on the modulation scheme, such as BPSK, QPSK, 16-QAM, and 64-QAM.

Much like other IEEE 802.11 standards, the 802.11a extension uses a version of CSMA/CA. A mobile client first senses the medium for a specific time interval and can initiate transmission if the medium is idle. Otherwise the client defers transmission and implements a backoff algorithm. Once the time period selected by the algorithm expires, the client can again attempt access to the medium. Much as in the basic 802.11 standard, RTS, CTS, and ACK frames are used to gain access to the medium and acknowledge that a frame was successfully received.

Because IEEE 802.11b products have attained a considerable level of acceptance and 802.11a products cannot interoperate with the 802.11b devices, the widespread use of the higher speed standard may not occur for quite some time. When products appear, their cost as well as the true throughput obtainable under different environmental conditions will govern their degree of adoption.

Installing a Wireless LAN

So far, we have primarily focused our attention on the theoretical aspects and generic hardware associated with wireless LANs. We can view that information as providing a foundation for this chapter which is, as its title implies, oriented towards the practical aspect of installing equipment that enables wireless LAN operations. In this chapter we focus our attention on the installation of different wireless LAN components that allow us to create a wireless LAN. We install a wireless LAN router that includes a built-in access point and, to make life interesting, wireless LAN network adapter cards from multiple vendors. The use of different client equipment provides a forum to describe and discuss interoperability issues. With that in mind, let us turn our attention to installing a product that was briefly described earlier in this book: the SMC Networks Barricade wireless broadband router.

The SMC Networks Barricade Router

The SMC Networks Barricade wireless broadband router is a multifunction device whose wireless operations comply with the IEEE 802.11b specification. As such, the Barricade supports direct sequence spread spectrum (DSSS) communications at 11 Mbps, 5.5 Mbps, 2 Mbps, and 1 Mbps over the air.

Product Overview

It is always a good idea to get an overview of the features, functionality, and capability of a product prior to its installation.

The Barricade broadband router contains three 10/100-Mbps Ethernet ports and one WAN port. The WAN port represents a conventional Ethernet interface that provides a connection to a cable or DSL modem or an Ethernet port on a conventional router, whereas the three Ethernet ports actually represent a three-port autosensing Ethernet switch. Through the use of one or more of those switch ports you can connect individual computers via an Ethernet connector or a network. Concerning the latter, you could cable individual hub ports to one or more of the three Ethernet ports, permitting up to 252 distinct IP addressable devices to share the resources of the router. The reason only 252 instead of 254 IP addresses are available is a result of the manner in which the Barricade internally allocates IP addresses. The Barricade uses one RFC 1918 Class C address for itself and a second for its gateway feature. This leaves 252 addresses available for client stations, which should be sufficient for most organizations. Because you would more than likely purchase the Barricade to use its wireless LAN capability, if you also need to support a wired infrastructure you must determine the total number of wired and wireless clients that can be supported by the router.

Features

Features supported by the Barricade include its virtual private network (VPN) support, its ability to assign RFC 1918 Class C addresses dynamically, a firewall capability that permits certain types of pockets to be filtered, an access control capability that permits different access rights to be assigned to clients, and virtual server support. Here the term *virtual server* is used by SMC Networks to denote the exposure of certain IP services, such as Web, FTP, and Telnet on an organizational LAN by Internet users. Other features supported by the Barricade include the ability to define attributes to support applications that require multiple connections (ports), which is referred to as "user-definable application sensing tunnel" and a Web-based configuration capability. We will extensively examine the Web-based configuration capability of the Barricade in this section to illustrate its installation setup, but prior to doing so, a few words are in order concerning equipment features.

Because the requirements of organizations can differ considerably, there is no one set of common features all organizations can be expected to use. Instead, organizations must evaluate different product features against their requirements, which often means that many product features may not be used. Because readers have a varied background, we will discuss as many of the major features of the Barricade and other products as is practical in this chapter. And because we assume a decision was made to use the Barricade, we will dispense with an evaluation of product features and commence the installation process with a discussion of a site location plan.

Site Location

As a wireless access point the Barricade is very similar to other access points concerning the selection of an applicable location to install the device. It differs from a conventional access point in that it not only provides connectivity to a LAN, but contains four wireable ports whose use must be considered along with the power requirements of the device and its wireless transmission capability. Thus, it is quite possible you may need to make a trade-off between positioning the Barricade for optimum wireless transmission capability and positioning it near a power outlet or directly connected PC and wired network connectors.

Wireless Positioning

For optimum wireless performance, the Barricade is no different from any access point. You should attempt to position the device in the center of the area where wireless clients will be located. In doing so, take into consideration any obstructions between the location of the Barricade's dual antennas and the wireless clients it is intended to serve, as well as possible sources of interference. Concerning obstructions, I was able to obtain a data transmission rate of 11 Mbps between the Barricade and a client within a radius of 100 feet. This 100-foot radius included transmission through the wood doors and floors in my home. However, when I took my trusty laptop outdoors, the transmission rate significantly declined as I moved further and further into my back yard. Because my back yard slopes away from my house, this was not a particularly good test of the range of the Barricade. Because SMC Networks provides a client utility program that monitors signal strength, I recommend that you initially install the Barricade and after powering it on, install an adapter card in a laptop and move about your facility to check signal strength to the locations you anticipate serving. As you move your laptop, you might also consider adjusting the physical position of the Barricade and its dual antennas. Because the Barricade can be mounted on a wall or on a flat surface you may be required to perform an iterative process until you select an optimum wireless connection location. As you perform this process you must remember that you have to consider both power cabling and any wired LAN or individual PC connectivity constraints.

Connectivity Tradeoffs

Figure 7.1 illustrates the connectivity tradeoffs associated with the installation of the Barricade broadband wireless router. The figure indicates that you must consider the ability of the router to support both wired and wireless clients as well as the power cable and Internet WAN connection. By considering each of these constraints, you will be able to select an appropriate location for the router. Although that location may not be optimum for all factors considered, at the very least a methodical positioning approach ensures that you consider the requirement of both wired and wireless clients.

Now that we understand the factors associated with the positioning of an access point that includes Internet and wired LAN support, we can configure the Barricade for operation.

Figure 7.1
Installing a Barricade
router with a built-in
access point
represents a
compromise
between wireless
and wired LAN
access as well as
connectivity to the
Internet and
availability of power.

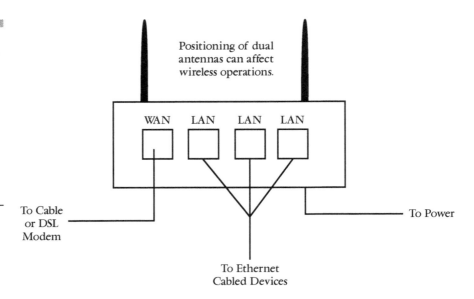

Figure 7.1
Installing a Barricade router with a built-in access point represents a compromise between wireless and wired LAN access as well as connectivity to the Internet and availability of power.

Using WINIPCFG

Prior to configuring any network device, it is a good idea to use WINIPCFG, a utility program included with most versions of Microsoft Windows. WINIPCFG provides a summary of the existing IP configuration of a computer.

The rationale behind the use of WINIPCFG is that it displays many key settings associated with an existing network. Even if you only have a PC with an Ethernet network adapter card, the use of WINIPCFG will, at a minimum, inform you of the layer 2 media access control (MAC) adapter address burned into read-only memory on the card. Although the setup of the Barricade router does not require this information, some other products do. In addition, it is advisable to use WINIPCFG to document your existing settings prior to installing any hardware that may require you to change your computer or network settings.

Figure 7.2 illustrates the display of the IP Configuration dialog box resulting from entering the command WINIPCFG in the Start>Run box. If you carefully examine the dialog box shown in

Figure 7.2, you note the absence of a print menu. Thus, to obtain a listing of the host and adapter settings you either need to put pen to paper or use a screen capture program to capture and print the dialog box.

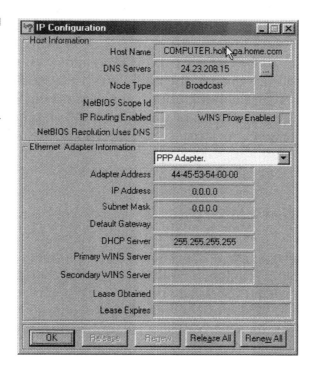

Figure 7.2
Viewing the IP Configuration dialog box generated by the WINIPCFG command.

The elements shown in Figure 7.2 resulted from the connection of my IBM NetVista computer to a cable modem. Because the cable modem and cable connections reside on the top of a desk in my kitchen, the cable connection was convenient for the cable installer, but access to the Internet was restrictive with respect to several key areas within and outside my home. Thus, an obvious solution was to use wireless technology instead of drilling holes in walls and floors in an attempt to extend coaxial cable to other areas of the home.

A second reason for the selection of wireless technology over conventional cabling was economics. In addition to being time consuming, if additional cabling was performed by yours truly, each connection point in my home would require either an additional cable modem if coaxial cable was routed to additional locations, or a wired LAN and router if category 3 or 5 twisted pair wire was routed to additional locations. Furthermore, once such cabling was completed it would not be a flexible solution, since on a periodic basis my wife or I rearrange our two home offices. In addition, the thought of crawling around the crawl space as well as under two decks to add a wired LAN connectivity capability outside the home was not appealing. Thus, even if I were writing a book about deep-sea diving instead of wireless LAN, the installation of a wireless LAN was the practical solution with respect to time, cost, and flexibility of the location of equipment.

Software Setup

Like most types of communications products, the SMC Networks Barricade Wireless broadband router comes preconfigured with certain default settings. This device has the default IP address of 192.168.123.254 and the default subnet mask of 255.255.255.0. For those of us a bit rusty with remembering IP addresses, the 192.168.123.0 network represents an RFC 1918 network address. RFC 1918 lists class A, B, and C network addresses that can be used for private Internets. Because the network addresses defined in RFC 1918 can be used by multiple organizations they are not unique and cannot be directly used on the Internet. Because the Barricade supports DCHP, however, it permits you to assign either dynamically or manually, 192.168.123.0 network addresses to devices that will use the Barricade to access the Internet. The Barricade performs an address translation, mapping 192.168.23.0 network addresses into a single IP address assigned to your Internet connection. To accomplish this, the Barricade performs network address translation by mapping 192.168.123.0 addresses into port numbers used by the single IP address assigned to your Internet connection.

Verifying Computer–Router Connectivity

Once you power up the Barricade and directly cable a PC to one of its Ethernet ports or install a wireless card in a client, you should verify its operational state. To do so, you can use the TCP/IP Ping utility program. Use the default IP address of the Barricade in the Ping command.

An example of the use of the Ping utility program is illustrated in Figure 7.3. In examining Figure 7.3, note that the Ping command operates as an MS DOS command and can be directly entered in the MS DOS Prompt dialog box. When used under Windows, Ping initiates the transmission of four Internet Control Message Protocol (ICMP) Echo Request Packets. When those packets are transmitted, the computer issuing the Echo packets also sets a timer. The target IP address, in this example the Barricade router, replies to each Echo Request, thus enabling the round trip delay from the client to the router to be computed. The occurrence of a reply indicates that a connection was established between the computer issuing the Ping and the Barricade. If, instead of the replies shown in Figure 7.3, your computer displays a series of "Request timed out" messages you

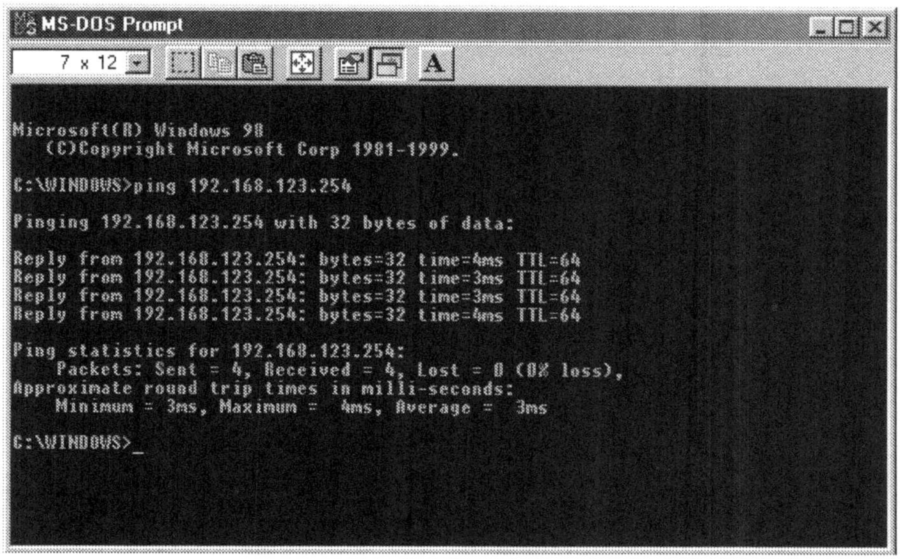

Figure 7.3
Using the Ping utility program to verify there is a connection between the router and a computer that will be used to configure the router.

```
MS-DOS Prompt                                                    _ □ ×

  7 x 12 ▾

Microsoft(R) Windows 98
   (C)Copyright Microsoft Corp 1981-1999.

C:\WINDOWS>ping 192.168.123.254

Pinging 192.168.123.254 with 32 bytes of data:

Reply from 192.168.123.254: bytes=32 time=4ms TTL=64
Reply from 192.168.123.254: bytes=32 time=3ms TTL=64
Reply from 192.168.123.254: bytes=32 time=3ms TTL=64
Reply from 192.168.123.254: bytes=32 time=4ms TTL=64

Ping statistics for 192.168.123.254:
    Packets: Sent = 4, Received = 4, Lost = 0 (0% loss),
Approximate round trip times in milli-seconds:
    Minimum = 3ms, Maximum =  4ms, Average =  3ms

C:\WINDOWS>_
```

do not have a valid connection to the router. In this situation, check that TCP/IP is installed on the computer you are using to configure the Barricade as well as the connection medium. This medium would be the Ethernet cable in a wired computer and a wireless LAN network card if you were configuring the router via a wireless connection. For either situation, you should check to ensure that the green-colored LED built into many network adapter cards is illuminated, or on some adapter cards flashing on and off. Assuming you made a valid connection, you can continue to set the configuration options available on the Barricade.

Configuring the Router

As mentioned earlier, one of the features of the Barricade is its ability to be configured through the use of a browser. If you use a PC that was previously configured as a LAN client and used a proxy service, you should disable the proxy and enter the router's IP address to access the main configuration window. This window, which is illustrated in Figure 7.4, displays the status of the router as well as providing the administrator with the ability to log into the router's operating system.

If you examine the top portion of Figure 7.4, you note the display of system status information for the WAN connection from the router to the Internet via my cable modem. This information informs us of the IP address and subnet mask issued by the Internet Service Provider (ISP), the ISP gateway's address, and the IP addresses of two Domain Name Servers (DNS) used for host-to-IP address resolution. In the lower portion of Figure 7.4, note the peripheral status section. Another feature of the Barricade is its shared printer capability. The router includes a printer port, which allows users to access a common printer, a capability that only a few years ago cost three times the total cost of the router.

If you look at the left portion of Figure 7.4, you note the software prompts for the entry of "admin" to log into the system. This is the default setting of the Barricade and should be changed at your earliest convenience.

Figure 7.4
The initial SMC
Networks Barricade
router screen display
shows the status of
the WAN connection
and informs you of
the default system
password.

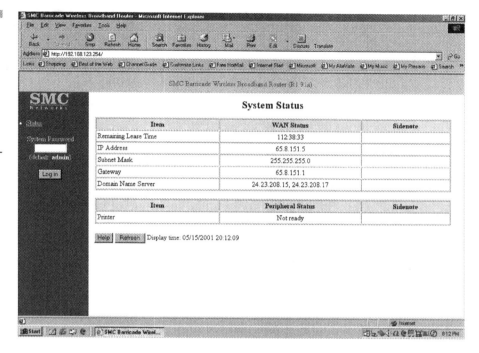

Figure 7.4
The initial SMC
Networks Barricade
router screen display
shows the status of
the WAN connection
and informs you of
the default system
password.

Configuration Options

Once you enter an appropriate system password, you have the
ability to select a number of configuration options. Whn you log
into the router your screen display changes slightly. Those
changes, which are shown in Figure 7.5, include the display of a
list of selectable configuration options in the left portion of the
screen and the ability to renew or release a dynamic IP address if
your ISP previously assigned a dynamic IP address using DHCP.

THE ADMINISTRATOR'S TOOLBOX. The second option in
the list of options shown on the left side of Figure 7.5 is titled "Toolbox." This option provides you with the ability to change the
default password.

Figure 7.6 illustrates the Administrator's Toolbox screen display.
Note that this screen is subdivided into two parts. The top portion of the toolbox permits the administrator or the first person

Figure 7.5
Once you log onto
the SMC Barricade
router you have
the capability to
configure the router.

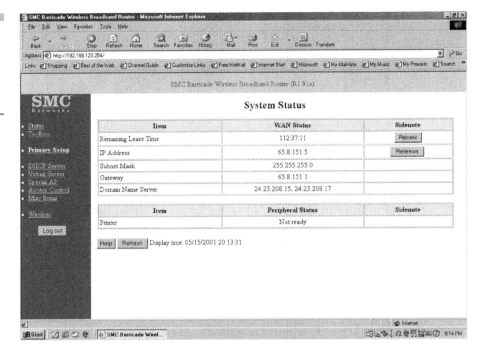

Figure 7.6
The Barricade's
Administrator's
Toolbox provides
users with the ability
to change the system
password as well
as to perform six
additional
operations.

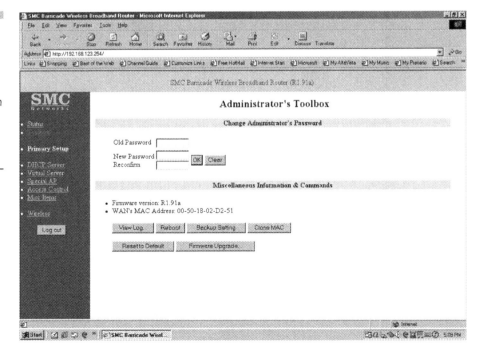

to log onto the Barricade to change its default password. Even if an unauthorized person should gain access to the router, the administrator (who should have physical contact with the router) can always perform a hardware reset to restore the device to its default settings and regain control of the Barricade.

The lower portion of the toolbox contains five buttons which, when pressed, perform the indicated operations. Although each button can be considered to perform an important operation, one that deserves attention is the button labeled "Firmware Upgrade." Periodically, SMC Networks updates the operating system of the Barricade, and you can obtain the most recent version of software from the vendor's Web site, thus permitting you to upgrade the router.

PRIMARY SETUP.   Continuing our exploration concerning the configuration of the SMC Barricade, let us turn our attention to the Primary Setup screen, which is displayed in Figure 7.7. This is one of the most critical screens because it is used to select the specific type of WAN connection interfaced to the router.

Figure 7.7
The Barricade Primary Setup screen permits you to change the router's default IP address and WAN setting as well as enter the host name assigned to your Internet connection.

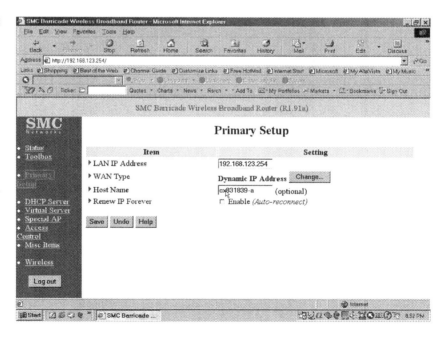

By default, the Primary Setup screen uses the LAN IP address of 192.168.123.254 for the router. The default WAN type is set to "Dynamic IP address" and presumes that your ISP uses DHCP to lease subscribers a single IP address. Although most cable and DSL operators use DHCP, not all do so. Thus, it is possible that you may have been assigned a static IP address, which will require you to select the button labeled "Change," whose effect we will shortly examine. In Figure 7.7, although the entry for host name is noted as "Optional," from my experience you need to complete this entry if you are using a cable modem.

When you click on the button labeled "Change" your screen display changes, enabling you to select one of four specific types of WAN connections. As indicated in Figure 7.8, you can select a static or dynamic IP address, the use of Point-to-Point Protocol (PPP) over Ethernet or dial-up networking, with the latter used to access the Internet via an ISDN or analog modem connection.

Figure 7.8
The SMC Barricade router supports four types of Internet WAN connections.

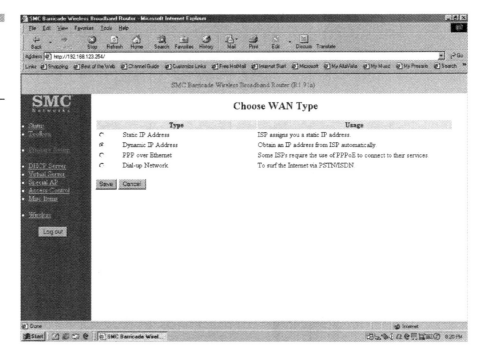

When you select an applicable WAN type, you are prompted to enter applicable configuration data. For example, if you selected a Static IP address you must enter the WAN IP address, subnet mask, gateway, primary, and if provided, secondary DNS addresses provided by your ISP.

DHCP SERVER. To facilitate the configuration of both wired and wireless clients supported by the Barricade wireless broadband router, this device includes a DHCP server capability. As indicated in Figure 7.9, the default setting of the DHCP server is enabled and 100 IP addresses are available, commencing at 192.168.123.100 and ending at 192.168.123.199. This represents a default address pool of 100 Class C addresses.

Figure 7.9
The DHCP Server screen provides users with the ability to enable or disable the dynamic host configuration protocol as well as allocate a block of IP addresses for automatic distribution to clients.

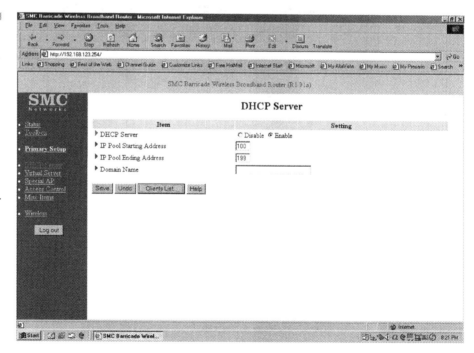

Whenever a DHCP client requests an IP address from the router, it allocates an unused address from the address pool. Although 100 IP addresses should be more than sufficient for most organizations, you can use addresses from 192.168.123.2 through 192.168.123.253. The default pool of 100 addresses allows users to use other 192.168.123.0 network addresses for special functions, such as a virtual server.

One of the features of the Barricade that some administrators learn to appreciate is its listing capability. One such list is a client's list of assigned IP addresses. This list is displayed by clicking on the button labeled "Client's list," shown in the middle of Figure 7.9. Doing so results in the display of currently assigned IP addresses as illustrated in Figure 7.10. Through the use of this list display screen during the busy hour, on a daily basis, you can determine if you need to adjust the number of IP addresses assigned to the DHCP address pool.

Figure 7.10
The display of the DHCP Clients list can be used to determine adjustments to the number of IP addresses assigned to the DHCP address pool.

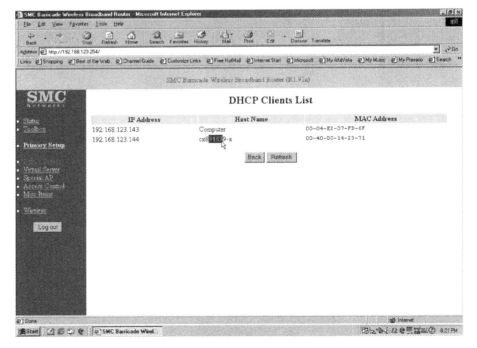

VIRTUAL SERVER. Because the Barricade router includes a network address translation capability, RFC 1918 Class C addresses are in effect invisible to the Internet community. The Barricade includes a feature referred to as *virtual server mapping,* which can make selected addresses accessible. To do so, you select the Virtual Server option from the left sidebar, resulting in the display shown in Figure 7.11.

Figure 7.11
Through the use of the Barricade's virtual server screen you can associate a well-known TCP/IP service port to an RFC 1918 address, permitting all requests for a specific service to be redirected to the indicated address.

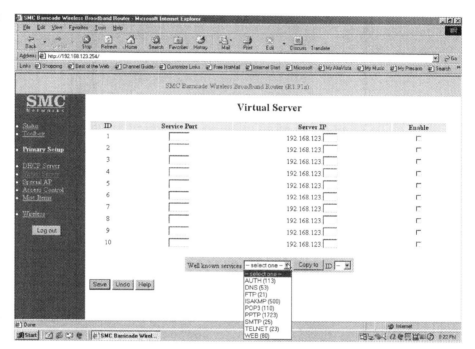

A virtual server represents a service port, where all requests to the specified port are redirected to the computer specified by an assigned IP address from one of the RFC 1918 addresses supported by the router. In examining Figure 7.11, note that you can select from nine supported services, such as FTP on port 21, Telnet on port 23, SMTP on port 25, and so on. In addition, you can also indicate another service by port number. For example, if you have an FTP server at 192.168.123.4 you enter 21 for the service port and 4 as the suffix for the address. Once this action is accomplished you click on the Enable box.

SPECIAL APPLICATIONS. Under the title "Special Applications" you can select a screen display to overcome one of the constraints associated with the basic network address translation (NAT) process. That constraint is associated with the use of certain applications that require multiple connections. For example, some Internet games typically use one predefined port for call establishment and a second port for the actual reception of game updates. Under NAT, each RFC 1918 address assigned to a client behind the Barricade router is normally mapped to a single valid IP address, representing an address provided by the ISP. Although this technique works very well for single connection applications, it prevents the correct operation of those that require multiple connections. To alleviate this potential problem the Special Applications screen provides administrators with the ability to define a single special application for a client. Once this is defined, the router's software in effect creates a tunnel through the device to support the specified multiple connection.

Figure 7.12
The Special Applications screen provides a mechanism to define applications that require multiple port connections that otherwise would be blocked by the router.

The configuration screen provided by the Barricade for special applications is illustrated in Figure 7.12. In examining the entries in Figure 7.12, note that the trigger field is used to specify the outbound port number that the multiple connection application issues first. The incoming port field is used to specify the port numbers of inbound packets associated with the application. This action informs the firewall feature of the inbound ports router that should be allowed to pass when packets with those port settings occur in response to the specified trigger.

To facilitate the entry of data, the Barricade screen display includes a series of predefined settings. Those settings are displayed in the lower portion of Figure 7.12 You can select an application and click on the "Copy to" button to copy the predefined setting.

ACCESS CONTROL.   Continuing our configuration of the SMC Networks Barricade router, the next screen is the Access Control configuration screen. This screen, illustrated in Figure 7.13, provides built-in firewall filtering capability. In examining Figure 7.13, note that the function of the Access Control screen is to enable the administrator to assign different access rights to different users.

The assignment of access rights results in the router's performing *packet filtering*—sending packets with an access right of "Block" to the great bit bucket in the sky, while packets that are associated with the access right of "Allow" are allowed to pass through the firewall.

If you carefully examine the router access control screen shown in Figure 7.13, you note that it supports four groups. You should divide your user community into groups. Otherwise, you are restricted to defining a limited number of access rights if you attempt to do so on an individual IP address basis. Of the four groups, the groups labeled 1 through 3 are used to assign access rights to specific clients, whereas all other clients are members of the default group. In using the numbered groups, you identify clients by the host position of their RFC 1918 address. For example, because the default DHCP address poll issues IP addresses from 192.168.123.100 through 192.168.123.199, if you wanted to restrict all default clients to HTTP (Web browsing on port 80 and Telnet port 21) you would select "Allow" from the pull-down bar to change the default setting of "Block" for a numbered group. You then specify the IP addresses as 100—199 and the allowable ports as 21 and 80, entered as 21,80.

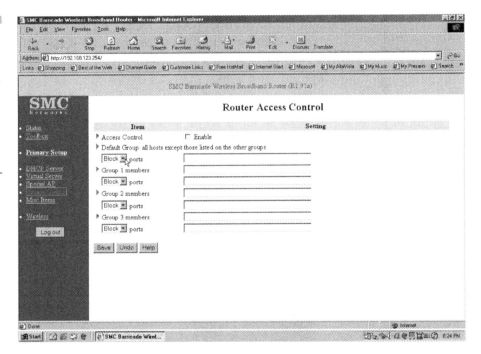

Figure 7.13
The Router Access Control screen provides the network administrator with the ability to select which client addresses and services to either block or allow.

When considering the use of the Barricade's router access control screen, it is important to note that, much as with a conventional router that uses access lists, your intentions do not take effect until you enable access control. Thus, after you configure your access control group or groups you need to click on the Enable box. In addition, you should click on the button labeled "Save" or your effort will disappear when power is removed from the router.

Now that we understand how packet filtering occurs, let us turn our attention to a screen that supplements some of the functions of access control and adds capability to the router.

MISCELLANEOUS ITEMS. The Miscellaneous Items screen is illustrated in Figure 7.14. This screen provides the Barricade router administrator with five optional settings that do not easily fit under a prior screen category. The first setting, "IP address of DMZ Host," provides you with the ability to specify one client behind the router that will have unrestricted two-way communications. If you have a public Web server to be connected to the Internet, you might consider specifying its address in this option.

Figure 7.14
The Miscellaneous Items screen provides the router administrator with the ability to set configuration parameters for five features that do not easily fit into other categories.

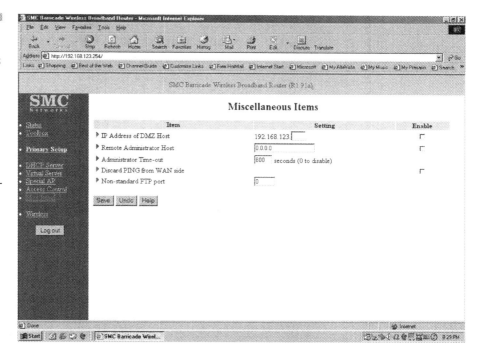

The second item on Figure 7.14, titled "Remote Administration Host," permits the administrator to access and control the router from a specific IP address. That address can be on your private network or even a host on the Internet. The default address of 0.0.0.0 allows any host to attempt to gain access to the router. Because of this, if you enable this option you should consider changing the address to a specific IP address. It should also be noted that, when enabled, the Web port is shifted from 80 to 88.

Although this action prevents unintentional access to your router, anyone with access to the manual or the SMC Networks Web site (as well as readers of this book) is aware of this port shift. Thus, you should probably use this feature sparingly and only with a specific IP address.

The third option, which is titled "Administrator Time-Out," by default disconnects an administrator after 600 seconds or 5 minutes of inactivity. As indicated in Figure 7.14, you can disable this feature by setting it to zero.

The fourth item, "Discard PING from WAN side," prevents ICMP Echo Request packets from flowing to the router. When this feature is enabled it prevents hosts on the WAN from pinging the Barricade.

The fifth item that you can set on the Miscellaneous Items screen is the port number, if you use a nonstandard FTP port. At one time it was common to use a nonstandard FTP port number to hide an FTP server from the general public.

Wireless Settings

In concluding our examination of the configuration of the SMC Networks Barricade broadband wireless router, we focus our attention on its wireless settings. While the title of the screen implies a single setting, in actuality several settings provide you with the ability to set certain wireless configuration parameters, including security.

Figure 7.15 illustrates the wireless setting screen display. This display includes three configurable parameters.

The first parameter you can configure is the network ID. The factory setting is default, which is sufficient if client stations do not roam between two or more access points. If clients roam between two or more access points, however, each access must have the same network ID.

The channel parameter represents the radio channel number. The factory setting is 6 for North America, 7 for Europe, and 7 for Japan. If you have multiple access points in close proximity to one another, each should be set to a different channel.

Figure 7.15
The Wireless Setting screen supports roaming as well as the enabling of wireless security.

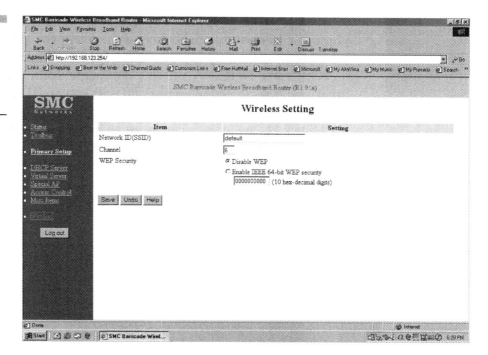

The WEP Security setting parameter is disabled by default. If you enable this setting, you must initialize the key used by all clients and the router. To do so you must enter a 10-digit hexadecimal code (0—9, A—F) in the router's security field and in each client's applicable security field.

Return to WINIPCFG

To conclude our examination of the SMC Barricade Wireless broadband router, we return to the use of the WINIPCFG utility program. This time we view the use of this program from the perspective of a client using wireless communications to the router.

Figure 7.16 illustrates the display of the IP Configuration dialog box after I entered the WINIPCFG command on my notebook computer, which is equipped with an SMC Networks EQ

Connect wireless LAN PC card. Note that the lower portion of Figure 7.16 contains Ethernet adapter information that shows that the notebook is using an RFC 1918 Class C address that was received via DHCP from the Barricade router. In addition, note that the default gateway and DHCP server IP addresses represent the default Barricade router IP address. Thus, from the use of WINIPCFG we can easily note that the computer on which Figure 7.16 was captured represents a client of the Barricade broadband wireless router. We can also note from the adapter that the computer in question obtains its connectivity via wireless transmission to the router.

Figure 7.16
Using the WINIPCFG program to view a wireless connection IP address assignment.

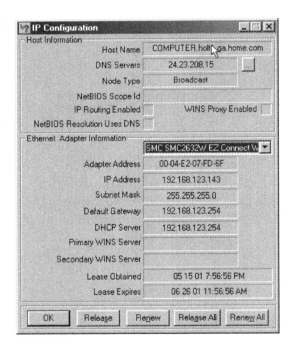

The SMC Networks EZ Connect PC Card

SMC Networks and other vendors offer wireless LAN adapter cards manufactured as PC cards and various types of ISA, EISA, and PCI adapter cards designed for insertion into the system expansion slot of a computer. Only the PC card form factor provides for true mobility, however, because the PC card is typically used in a laptop or notebook—which is obviously easier to move than a desktop computer. Another advantage of the use of a notebook or laptop computer that might be overlooked until you actually work in a wireless LAN environment is the ease of positioning portable computers to obtain an optimum level of signal strength. Because of the advantages associated with the use of PC card wireless LAN network adapters and the fact that I have six notebook computers and three desktops, I decided to illustrate the installation of a wireless LAN based on clients using wireless LAN network adapters fabricated as PC cards. Thus, in this section we turn our attention to the installation and operation of the SMC Networks EQ Connect wireless PC card. In doing so we briefly describe the driver installation, while primarily focusing our attention on several settings that can result in the more efficient operation of your over-the-air communications.

Driver Installation

SMC Networks wireless LAN PC cards and PCI adapter-based cards are similar because they are bundled with two diskettes. One diskette contains drivers for various versions of Microsoft's Windows operating system while the second diskette contains a wireless LAN utility program. The utility program gives you the ability to change the configuration of the wireless adapter after installation and to view the level of signal strength and signal quality in real time.

When you push the SMC Networks EZ Connect PC card into a type II slot, your computer's operating system immediately recognizes the additional hardware and a wizard prompts you through the driver installation process. After you enter the applicable location for the wizard to search for the driver, it copies files from the specified location on the diskette to your hard drive. If you are using Windows 2000, you must select Microsoft's TCP/IP protocol to use with the adapter,; this is accomplished automatically for other versions of Windows. Once the protocol is installed a dialog box is displayed that contains seven selections that can be used to tailor the configuration of the wireless LAN adapter card for a particular operational environment.

ADAPTER CONFIGURATION. Figure 7.17 illustrates the dialog box displayed when installing an SMC Network wireless LAN adapter card. As you highlight each of the seven options in the left window labeled "Property" you have the ability either to keep the default setting or change the value associated with the selected property. Because some of those properties require a bit of clarification, let us examine each.

Figure 7.17
The installation of the SMC Network EZ Connect Wireless LAN adapter requires setting seven properties.

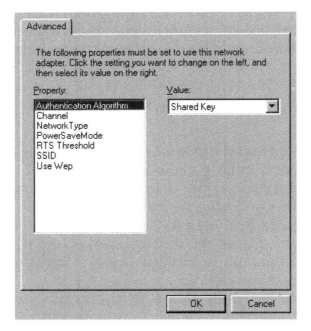

AUTHENTICATION. The authentication algorithm property governs the manner by which clients are authenticated. The default is set to *shared key*, which means that the two parties communicating use the same encryption key. Because encryption by default is disabled however, this also means that authentication is disabled. The optional value for the authentication algorithm is *Open System*. Products from some vendors use this setting, so you would select Open System to obtain compatibility with other vendor equipment that supports this setting.

CHANNEL.   The Channel property provides you with the ability to set up an ad hoc network under which two or more clients communicate with one another directly without the use of an access point. If you are communicating with an access point the adapter will automatically scan frequencies and set the channel to that used by the access point.

NETWORK TYPE.   The third property you can set is Network Type. By default this property is set to "Infrastructure," which is the setting required to obtain access to a wired LAN via an access point. The other two possible settings are ad hoc and 802.11 ad hoc. Both ad hoc and 802.11 ad hoc modes are peer-to-peer communications methods that bypass the access point.

POWER SAVE MODE.   The Power Save Mode parameter is set to *disabled* by default. If you are using battery power, however, you may wish to change this setting to *enabled* to reduce power loading.

RTS THRESHOLD.   The RTS Threshold setting represents the time a client waits prior to issuing an RTS frame. The default setting for the SMC Network EZ Connect adapter is 2432, which disables the threshold. When using a mixture of vendor equipment, you may need to change this default to the same value used by other devices in your network.

SSID.   The Service Set identification (SSID) represents the wireless network identification number. By default it is set to a value of "Any," which enables the client to use the identifier of an access point because the default network type is set to "Infrastructure." The SSID setting should be changed to the same identification used by other members of an ad hoc work group, or from a different vendor's access point that you need to connect to that does not support automatic recognition.

USE WEP. The last property you need to consider during the installation process is Wireless Equivalency Privacy (WEP), which by default is set to disabled. The other options supported by the SMC adapter card include 64-bit and 128-bit encryption. Selecting 64-bit encryption requires the entry of 10 hex digits that are added to a 24-bit vector to form a 64-bit key. Selecting 128-bit encryption requires the entry of 26 hex digits which when added to a 24-bit vector, result in a 128-bit key. We initially accept the default of WEP disabled; later in this section we will use the SMC-supplied utility program to change that setting.

Before examining the use of the utility program furnished with the adapter, a few words concerning property option selections are in order. Although there are seven properties you can consider, if you are accessing an SMC access point and did not change any equivalent properties on that device, you can forego selecting client properties and obtain an instant connection to the access point. That said, let us move on to discuss the operation and utilization of the utility program bundled with the wireless LAN adapter.

Configuration Utility

The configuration utility program bundled with the SMC wireless LAN adapter not only provides you with the ability to make after-installation configuration changes but provides signal quality and signal strength indications that are valuable in positioning equipment.

CONFIGURATION UTILITY ICON. Figure 7.18 illustrates the EZ Connect Wireless window that shows the icon for the configuration utility after it was installed on my notebook. The primary reason for the display of Figure 7.18 is to acquaint you with the shape of the configuration utility icon, which resembles a PC with an antenna radiating what appears to be two radio waves.

Figure 7.18
The Configuration Utility icon resembles a PC with an antenna radiating two radio waves.

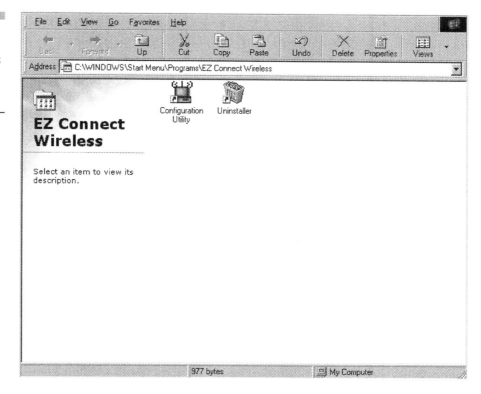

The icon shown in the left portion of Figure 7.18 will also be installed on the task bar next to the time display. When colored green, it indicates that your adapter card has an over-the-air connection; when colored red, it indicates the absence of a connection. You can either go to Start>Programs>PRISM 802.11 Wireless LAN>Configuration Utility or simply click on the funny looking icon on the task bar to invoke the configuration utility.

LINK QUALITY.   Figure 7.19 illustrates the display of the Wireless LAN configuration utility in the center of my notebook screen. If you look at the lower right corner of the screen, you see the icon. Double clicking on that icon causes the display of the utility screen, with its Link Info tab positioned in the foreground. If you carefully examine Figure 7.19, you also note that Channel 6 is being used and the current transmission rate is only 2 Mbps. Also note that the link quality and signal strength are shown as good, with the link quality at 80 percent and the signal strength at 73 percent.

Figure 7.19
Viewing the Link Info tab on the wireless LAN Configuration Utility screen.

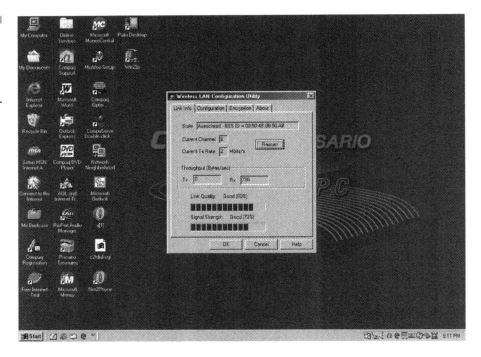

Previously I noted that the Link Info tab can be used to position a client to obtain an optimum level of transmission. To illustrate this positioning capability, I moved my notebook so that the wireless adapter card's antenna was positioned directly toward the antenna of the Barricade router. In addition, I clicked on the button labeled "Reset." The result of these two operations is shown in Figure 7.20. At this point, my link quality and signal strength had significantly improved and the transmission rate increased to 11 Mbps.

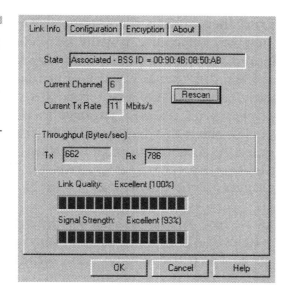

Figure 7.20
Viewing the effect of repositioning the antenna of the wireless LAN adapter card and initiating a channel rescan.

SETTING WEP. The second tab in the configuration utility program provides you with the ability to set the network mode, SSID, transmit rate, power share mode, and WEP. When you change the WEP setting from its default of disabled, you have the ability to use the tab labeled "Encryption."

Figure 7.21 illustrates the Wireless LAN configuration utility display with its encryption tab in the foreground. Normally the entries in this tab are shaded gray because the default setting for WEP is disabled, but if you set WEP, you have the ability to use the encryption tab.

The SMC Network EZ Connect Wireless LAN adapter card's driver supports the creation of up to four encryption keys, of which only one can be used at any one time. Although many software products just prompt the user to enter a 10-digit hex code, one of the more interesting aspects of the SMC Network utility is its support for the use of either the entry of a pass phrase or a sequence of hex digits to create a key.

In Figure 7.21, I entered the time honored phrase "nowis-thetimetoeatchicken." Clicking on the button labeled "Generate" resulted in the generation of four groups of 10 hex digits, with the default key 1 set to hex 8d76a34ee3. Clicking on the button labeled "Write" then updates the adapter card's driver and registry with the new key. Of course, if your LAN card is in infrastructure mode, and your access point maintains the default of WEP disabled or has a different key, you will not be able to communicate. Similarly, if your LAN adapter card is in ad hoc network mode of operation and you attempt to communicate with another wireless client that has either WEP disabled or a different key, you can also forget about communicating until you have compatible settings. Unlike Danny DeVito and Arnold Schwartzenegger in the movie *Twins*, your WEP keys have to be identical.

Now that we have an appreciation for the manner by which we set up a wireless LAN adapter to communicate with an access point, let us go a step further and turn our attention to communications using different vendor equipment.

Figure 7.21
The SMC Network Configuration utility program permits encryption keys to be generated either by the entry of a pass phrase or through the entry of a sequence of hex digits.

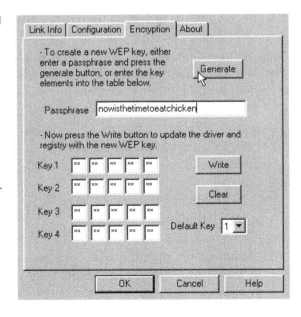

Agere Systems Orinoco PC Card

This section focuses on the use of an Agere System Orinoco PC card. For those readers not familiar with the reorganization of corporate America, Agere Systems was formerly part of Lucent Technologies and became an independent company during 2001.

Installation

The installation of the Orinoco PC card is a bit backwards from the manner in which most other hardware products are installed. Instead of first installing the hardware and using the Microsoft Wizard to install the applicable driver, you first use the provided CD-ROM to install the Orinoco PC card software. Although this installation reversal is well documented in the Orinoco Getting Started guide, if you "jump the gun" and insert the hardware first you might encounter a problem. At a minimum, you will be confused about what to do next and you may have to read the instructions.

Figure 7.22 illustrates the initial screen display generated by the CD-ROM that accompanies the PC card. Once you install the software, you must insert the PC card into a Type II card slot. This action activates the Microsoft wizard and allows you to install the applicable driver.

The utility program installed from the CD-ROM can be activated after the driver is installed by one of several methods, similar to the manner by which the SMC Network configuration utility program is invoked. You can activate the program by double clicking on a signal strength indicator that is added to the taskbar near the time display or from the Start>Programs>Orinoco sequence. Once we complete our description of the installation of the driver and the initial configuration of the Orinoco PC card, we will examine the use of the utility program in detail.

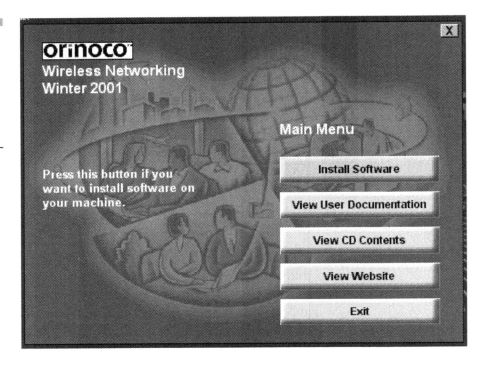

The installation of the driver program results in the generation
of a dialog box labeled "Add/Edit Configuration Profile," which
enables you to tailor the PC card's parameters. This dialog box is
illustrated in Figure 7.23. In examining Figure 7.23. note that the
Orinoco PC card software permits you to define four profiles,
allowing you to switch between profiles as you move from office
to home or to another location. The name of the default profile is
"Default," and it is shown configured to provide access to an access
point. Through the pull-down tab you can also select "Residential
Gateway" or "Peer-to-Peer Group."

PROFILE OPTIONS.   By clicking on the button labeled "Edit
Profile" in Figure 7.23 you have the ability to tailor a profile to
your specific networking requirements. When you click on Edit
Profile, the label of the dialog box changes to "Edit Configuration"
and four tabs are displayed. Figure 7.24 illustrates the Edit Configuration dialog box with its Basic tab in the foreground. Note that

Figure 7.23
The Orinoco Add/Edit Configuration utility provides up to four profiles to be predefined.

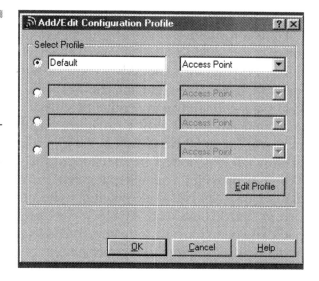

the only option on this tab is for the entry of a network name. If you are communicating with an Orinoco residential gateway access point, you must enter the ID number printed at the back of the access point. Because I was attempting to illustrate interoperability between the Agere System PC card and an SMC Network Barricade router, no entry was initially entered into the network

Figure 7.24
When accessing an Orinoco access point you must enter the ID of the access point. Otherwise, a network name of ANY should be entered to access a third-party product.

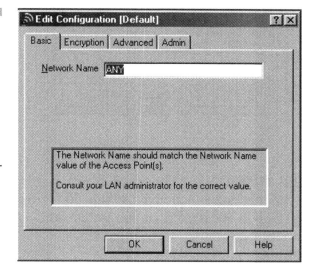

name field. Unfortunately, the omission of the keyword ANY was a mistake, as it took several minutes to figure out why the Orinoco PC card showed a good level of signal strength but was incapable of communicating with the SMC Network Barricade router. Once the keyword ANY was used for the network name, I was able to get interoperability between vendor products.

The second tab in the Edit Configuration dialog box is labeled "Encryption." The Edit Configuration dialog box with the Encryption tab positioned in the foreground is shown in Figure 7.25. This tab provides you with a mechanism to enable wireless security. As in SMC Network products, the default setting of the Orinoco PC card is no security, since the checkbox is blank. Note that you can specify up to four keys and use either alphanumeric or hexadecimal characters to form the encryption keys.

Figure 7.25
The Encryption tab in the Edit Configuration screen supports the use of either alphanumeric or hexadecimal characters to form up to four encryption keys.

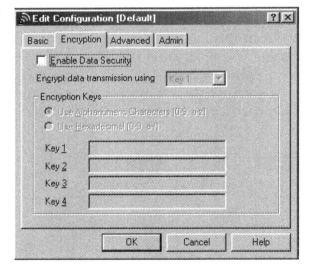

OTHER OPTIONS. The third tab shown in the Edit Configuration dialog box is labeled "Advanced." This tab controls three settings and is illustrated in the foreground of the dialog box in Figure 7.26.

Figure 7.26
The Advanced tab controls the settings associated with power management interface robustness and RTS/CTS medium reservation.

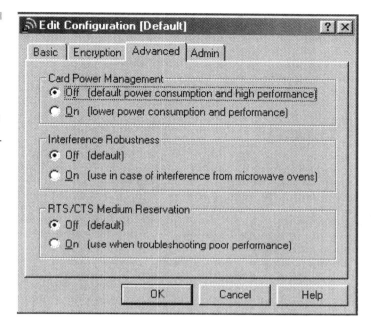

In examining Figure 7.26, note that by default, power management is disabled. If you use battery power, you may wish to consider turning this option on.

The second option on the Advanced tab governs the robustness of the interface. By default this option is off; however, if you are working in a kitchen or in an office near a microwave oven you may wish to turn this option on to minimize the potential interference of a microwave's magnetron that generates RF radiation in the 2.4-GHz band.

The third option on the Advanced tab controls the use of RTS/CTS packets to reserve the medium. By default this option is also turned off.

ADMINISTRATION. The fourth tab in the Edit Configuration dialog box is the Admin tab. Figure 7.27 illustrates the positioning of this tab in the foreground of the dialog box.

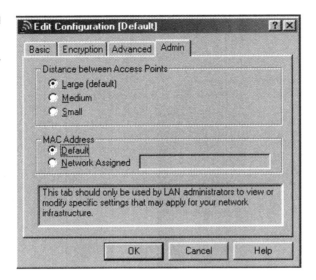

There are two options on the Admin dialog box. The first option permits you to specify the distance between the client and access point. This option controls the roaming sensitivity of client stations and should be set to match the setting of an Orinoco access point. If you are accessing a different vendor's access point it is more than likely there in no similar setting and you should not vary this setting from its default value.

The second option provides you with the ability to define a MAC address. The default setting uses the universal MAC address contained in read-only form on the PC card. If your organization uses local addressing, commonly referred to as *local administrated addressing,* you click on the button next to "Network Assigned" and enter the local address. As noted by the information on the screen, this tab is for use by a network administrator, the person responsible for controlling MAC addresses when local addressing is used.

The Client Manager

The Orinoco Client Manager gives you the ability to edit the configuration profile created during the installation process. In addition, through the use of the Client Manager you can initiate a diagnostic test of the PC card.

Figure 7.28 illustrates the display of the Orinoco Client Manager. Note that this dialog box informs you of the signal strength and network status of your connection. The File menu option permits you to either disable the radio link or close the window. The second menu, Actions, allows you to change a previously configured profile or add a new profile. The third menu, Advanced, gives you the ability to perform several types of diagnostic tests. Those tests include card diagnostics as well as a link test.

Figure 7.28
The Orinoco Client Manager provides the ability to add or edit configuration profiles and perform different diagnostic tests.

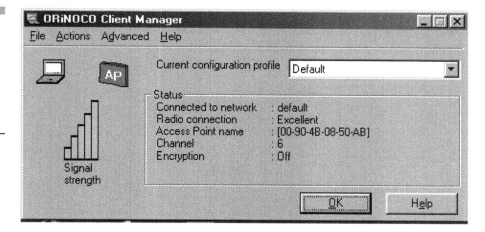

In examining Figure 7.28, note that the status information informs us that we have an excellent radio connection and the name of the access point is 00-90-4B-08-50-AB. That is the name of the MAC address of the SMC Network wireless broadband router I use to surf the Internet. Also note from the status information that we are using Channel 6 and encryption is off.

If you focus on the left portion of Figure 7.28, you note a series of five ever-increasing vertical bars labeled "Signal Strength." This symbol is also displayed as an icon on the taskbar, which you can double click to activate the client manager.

DIAGNOSTIC TESTING.   Selecting Card Diagnostics from the Advanced menu shown in Figure 7.28 provides you with the ability to check the status of the Orinoco PC card. Selecting that option results in the screen shown in Figure 7.29. Note that the error details entry informs us that, although the radio link is OK, the card was not able to retrieve the access point name. This error message results from the fact that we are accessing an SMC Network access point that neither requires nor uses a name.

Figure 7.29

Through the diagnostic capability of the Orinoco Client Manager you can test the PC card.

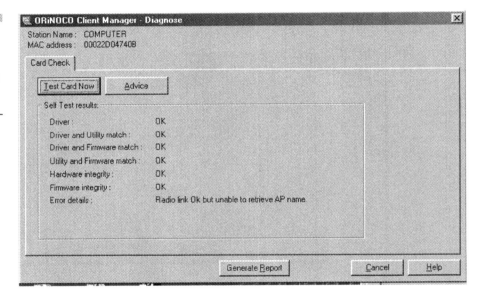

Proof of the Pudding

A favorite expression of one of my teachers was "the proof of the pudding is in the eating." Thus, in concluding this chapter, we test the use of the Orinoco Client Manager superimposed on the use of Microsoft's Internet Explorer to access the Internet, verifying that a connection to the Internet was accomplished.

Figure 7.30 illustrates the screen display on my computer using an Orinoco PC card to access the Internet via an SMC Network

Barricade wireless broadband router connected to my cable modem. In examining Figure 7.30, note the signal strength indicator shown in the Client Manager is also displayed as an icon on the task bar in the lower right portion of the screen. By simply double-clicking on the icon on the task bar, you can invoke the display of the Client Manager.

Figure 7.30
Wireless access to the Internet via an Orinoco PC card communicating with an SMC Network access point.

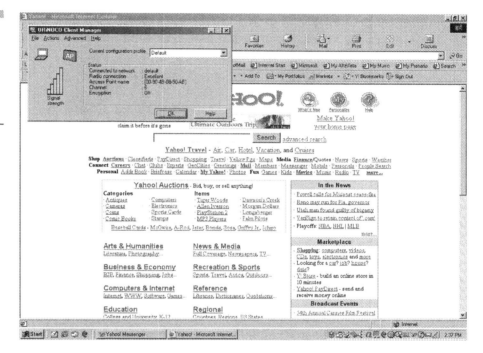

Although access to the Internet is shown in Figure 7.30, a few words concerning Internet Explorer settings are warranted. To access the Internet, several changes to the configuration of the Internet Explorer were required. First, I had to go to the Tools menu and select Internet Options and then select Connections. Because I previously used a dial-up connection to the Internet, that connection method had to be revised to enable Internet Explorer to use a wireless LAN connection.

A second change that required a few minutes of effort involved adding the entry ANY to the network name dialog box. Before this was done, I could see the connection to the access point but could not communicate with the device. Once ANY was added as the network name to the Orinoco PC card configuration, the ability to access the Internet immediately occurred.

Although it took me a few hours to write a description of the installation and operation of the Orinoco PC, card in actuality the process required 10 to 15 minutes to modify the configuration of the Internet Explorer to illustrate wireless interoperability. Thus, in concluding this chapter we can note that IEEE 802.11b interoperability is a fact as long as you are willing to tinker with configuration settings.

The Home
RF Standard

In this chapter we examine an evolving wireless specification tailored for use in the home and small office. Considered by some to represent competition to the IEEE 802.11 series of wireless LAN standards, in actuality it is probably a bit soon to decide if the Home RF standard represents a product category competitive with IEEE LANs. Only recently did the Federal Communications Commission (FCC) allocate additional bandwidth of 3 and 5 MHz for use by frequency hopping spread spectrum (FHSS) operations in the 2.4-GHz band. This increase enables Home RF-compliant products to operate at 10 Mbps, which will be more competitive with the IEEE 802.11b standard than the first generation of Home RF products that operated at 1Mbps. Because Home RF products operating at 10 Mbps will be reaching the market at the time this book is published, it may be a while before we can determine the effect of one technology on the other. The old adage that "competition promotes discounts," however, can be expected to lower the cost of both Home RF and IEEE 802.11 products due to the large number of vendors manufacturing devices compliant with one or both standards. The second section of this chapter provides technical insight into the Home RF standard, including a detailed description of many of its features.

Overview

We use the term *wireless home networking* to represent the use of the radio frequency spectrum to transmit both voice and data in the home or in a small office environment. The basis for wireless home networking is the Home RF Working Group's Shared Wireless Access Protocol (SWAP).

Wireless Home Networking represents a technology developed by an industry consortium known as the Home RF Working Group. Because of the name of the working group many publications refer to Wireless Home Networking as Home RF, although technically the term refers to an industry consortium.

Home RF products are designed with a range of up to approximately 150 feet, which permits full coverage within all homes except mansions. The technology uses the 2.4-GHz ISM band, which allows transmission to occur through most home barriers, such as conventional floors and walls.

The Home RF specification operates at the Physical and Data Link layers of the OSI Reference Model. This enables the specification to function as a transport mechanism for both voice and data, as we will shortly note. In fact, a key difference between the Home RF specification and IEEE 802.11 wireless LANs is the fact that the former supports the Digital Enhanced Cordless Telephone (DECT) cordless standard that allows Home RF equipment to support both voice and data.

At the time this book was prepared, eight vendors had announced products based on the Home RF Working Group's Shared Wireless Access Protocol (SWAP). SWAP represents both the actual wireless protocol used to convey information among Home RF-compatible devices and its series of specifications. Vendors announcing Home RF products as this book goes to press include Cayman Systems, Compaq Computer, IBM, Intel, Motorola, Scientific Atlanta, Pace, and Proxim.

Versions

Home RF was originally designed as a mechanism to provide wireless broadband communications for the home. Since the original development of the Home RF SWAP specification, several minor and major revisions to the standard have occurred or are expected to occur in the near future.

Home RF Version 1.0 provided a data rate of 1 and 2 Mbps, similar to the original IEEE 802.11 standard. Home RF 2.0 extends transmission to 5Mbps and 10Mbps via the use of 3 MHz and 5 MHz of bandwidth, respectively. Home 2.0 products are backward compatible with the installed base of Home RF 1.0 devices. On the drawing board, Home RF 3.0, although at the time this book was written it lacked an official designation. This new version of Home RF is expected to extend data transmissions to 20 Mbps.

Network Architecture

The network architecture of a Home RF network is both similar to and different from the structure of an IEEE 802.11 network. Both the Home RF and IEEE 802.11 networks can operate either as ad hoc or managed networks. An ad hoc network permits one Home RF compliant device to talk with another without requiring the use of an intermediatory device. In a managed network, the Home RF network employs a connection point (CP) as the central transmission location or focal point.

THE CONNECTION POINT. The Home RF CP is similar to an IEEE 802.11 access point; however, there are some distinct differences between the two. Like an access point, the Home RF connection point provides a gateway service to the Internet via a DSL or cable modem connection. Unlike the access point, however, the connection point provides a gateway to the public switched telephone network, which enables the CP to support both voice and data services. In addition, the connection point can be used to support power management of remote devices by scheduling device wake-up and periodic polling of devices. This action results in an extension of the battery life of remote devices and has no counterpart in the wireless LAN world.

Nodes

The design of the Home RF specification enables a network to accommodate a maximum of 127 nodes. Nodes can include a connection point that supports both voice and data services, voice terminals that use time division multiple access (TDMA) services to communicate with a base station, data nodes that employ the carrier sense multiple access with collision avoidance (CSMA/CA) protocol to communicate with other nodes in an ad hoc manner or with a base station in a managed network environment, and a combined voice and data node. The latter can use both voice and data services supported by the Home RF standard.

System Requirements

In a manner similar to that of the IEEE 802.11 wireless LAN specification, many Home RF wireless devices are fabricated as PC cards designed for insertion into a Type II PC card slot. That type of slot is included in just about all laptop and notebook computers manufactured after 1997. Because most desktop computers as well as printers and plotters that could be shared do not have card slots, Home RF devices are also fabricated as stand-alone devices with USB ports. This allows Home RF modules to be connected to a variety of home computers and peripheral devices that need to communicate with one another via the 2.4-GHz frequency band.

Technical Characteristics

The Home RF Shared Wireless Access Protocol (SWAP) is designed to support both voice and data as well as interoperate with the Public Switched Telephone Network (PSTN). In fact, the Home RF physical layer is implemented using components developed originally for the European Digitally Enhanced Cordless Telephone (DECT) standard. DECT represents the world's most successful cordless standard, with over 50 million products manufactured by over 100 certified vendors. If you visit a Kmart or Wal-Mart and gravitate to their electronics sections you may notice a large selection of cordless telephones. Those that have a sticker on the box labeled "2.4-GHz operation" are DECT-compliant phones.

FHSS Use

Much as in one of the three 802.11 access methods, the Home RF SWAP specification defines the use of frequency hopping, with a hopping rate of 50 hops per second. This means that the duration of each hop is 20 ms.

The use of frequency hopping occurs in each version of the Home RF specification, with the key difference between each version based on the amount of bandwidth available for use. For example, when 1MHz is available for use there are 75 channels over which frequency hopping can occur. When 5 MHz of bandwidth is available, frequency hopping can occur over 15 channels.

Power, Operating Rate, and Modulation

The transmission power of a Home RF system is 100 mW, whereas the obtainable data rate can be either 0.8 or 1.6 Mbps. At an operating rate of 0.8 Mbps, 2-FSK modulation is employed, whereas at an operating rate of 1.6 Mbps 4-FSK modulation is used.

Device Support

As previously noted, up to 127 devices or nodes can be supported on a common Home RF network. This should be more than sufficient for a home or small office environment. Each network is assigned a 48-bit network identification, which enables the concurrent operation of multiple collocated networks in the event a small business grows.

The use of a 48-bit network identification permits the Home RF specification to support the 6-byte addressing scheme used by IEEE 802 wired and wireless LAN nodes. Thus, if you dig down deep into the Home RF specification you can note many similarities between that specification and the work of the IEEE.

We previously mentioned that a Home RF network can support up to 127 devices. Those devices, which represent wireless nodes, can be a mixture of four basic types of devices described in Table 8.1. If you examine the entries in Table 8.1, it is obvious that the vision of Home RF is to provide interconnection capability for a wide range of devices. In fact, a Home RF system is capable of supporting up to eight full-duplex voice conversations as well as a mixture of computers and peripheral devices.

TABLE 8.1

Home RF Node
Types

Connection point that supports voice and data services.
Voice terminal that uses TDMA to communicate with a base station.
Data node that uses CSMA/CA protocol to communicate with a base station and other data nodes.
Voice and data node that can use both types of services.

Security

One of the key advantages of the Home RF standard in comparison to the IEEE 802.11 standard is in the area of security. Currently IEEE 802.11 wireless LANs use the Wired Equivalent Privacy (WEP) protocol, to provide both authentication and privacy equivalency. The 64-bit WEP key has been proved too short to prevent the compromise of network traffic. While a new IEEE standard, which will be described in Chapter 9, attempts to eliminate the authentication problem resulting from a shared key that can be compromised, to ensure privacy, 802.11 wireless LAN users must consider the use of 128-bit keys. While some vendor products support 128-bit keys, they do so in a proprietary manner: If you want to extend encryption, you need to first locate products that support a 128-bit key and then standardize on one vendor's product line. In comparison, the Home RF standard defines the use of a 128-bit encryption key. Thus, equipment that supports the standard also supports the 128-bit key. That means all standard compliant equipment will interoperate regardless of manufacturer.

The actual encryption algorithm specified by the Home RF SWAP system is the *Blowfish encryption algorithm*. This algorithm is a symmetric block cipher that was designed by Bruce Schneier of Counterpane Internet Security in 1993 as a drop-in replacement for the Data Encryption Standard (DES) algorithm. Unlike DES, which uses a 64-bit key, Blowfish supports a variable length key from 32 bits to 448 bits. The Home RF standard defines the use of a 128-bit key and the use of the Blowfish encryption algorithm, which results in over a trillion codes.

In addition to providing a higher level of security than WEP, Blowfish is an extremely fast algorithm and the overhead in encrypting data is minimal in comparison to the use of other encryption algorithms.

Data Compression

One of the more interesting features of the Home RF standard is its use of the *LZRW3-A data compression algorithm*. This compression algorithm represents a variant of the Lempel—Ziv—Ross Williams algorithm created by Dr. Ross Williams, which was based on the work of Professors Lempel and Ziv, who published two papers during the 1970s dealing with an algorithm for text compression that became the basis for a series of algorithms implemented in modems, operating systems, and software applications. Because of patent issues, several people including Dr. Williams, expanded on the work of Lempel and Ziv and created a string compression algorithm that bypassed patent problems and improved on the compression capability of the string manipulation process. The LZRW3-A algorithm requires only 16 K of memory and runs approximately twice as fast when compressing data and three times faster when decompressing data than the built-in UNIX operating system compression algorithm.

Home RF Operation

Earlier we noted that the Home RF specification supports both voice and data transmission. In this section we turn our attention to how the mixing of voice and data is accomplished by examining the lower two layers that represent the standard—the Physical and Media Access Control layers.

The Physical Layer

At the Physical layer, the Home RF standard uses frequency hopping spread spectrum (FHSS) communications in the 2.4-GHz unlicensed ISM band. This use of FHSS is uniform across each version of the Home RF standard, with version 1.0 using seventy-five 1-MHz channels to support data rates of 1 and 2 Mbps. In actuality, because of the design of the Home RF MAC layer, which is described in the next section, only a subset of the transmission time is available for data networking. As a result, the initial version of the Home RF standard supports effective data rates of 0.8 and 1.6 Mbps, whereas version 2.0 of the standard supports effective data rates of 5.0 and 10.0 Mbps as well as backward compatibility with the earlier standard.

Concerning version 2.0 of the standard, as previously noted, the ability to transmit at higher data rates results from the FCC permitting the use of 3 and 5 MHz of bandwidth in the 2.4-GHz frequency band. When 5 MHz is available, using FHSS the Home RF standard uses fifteen 5-MHz channels.

The MAC Layer

At the MAC layer the Home RF specification uses a combination of contention-based wireless Ethernet with priority access and time reservation through the use of time division multiple access (TDMA).

The Home RF MAC layer uses time slots to allocate data and voice calls. Figure 8.1 illustrates the allocation of the Home RF layer by time. In examining Figure 8.1, note that the bulk of the time slots are allocated to data networking. Within the time period allocated to data networking, streaming media sessions are given priority access.

Figure 8.1
The Home RF MAC
layer by time.

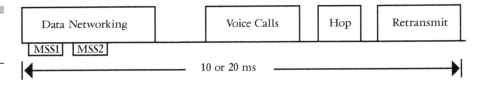

Legend:
MSS   Media Streaming Session

The actual period for voice calls is based on the number of active voice calls. If a packet transporting voice fails, it can be retransmitted at the start of the next frequency used by the frequency hopping sequence.

Although the retransmission of packets transporting voice is normally not considered in a Voice-over-IP environment because it can do more harm than good, in a Home RF environment the situation is different. The key difference that enables a failed packet to be retransmitted is the minimal amount of latency between packets on a Home RF network, because voice is transported in frames representing a 10 ms period. In comparison, the delay through an IP network can be hundreds of milliseconds in duration, which precludes the retransmission of a packet transporting a 10 or 20 ms snippet of a voice conversation. In a long delay environment, dropped or errored packets transporting voice are compensated for by the generation of noise, which works fairly well until too many packets are lost.

Frame Duration and Types

The time period allocated to a Home RF MAC frame is either 10 or 20 ms, with the period available depending on the presence of active voice calls. Up to three types of services are supported within the Home RF MAC layer. First, an asynchronous, connectionless packet is used to support the transport of TCP/IP traffic through the slot labeled Data Networking in Figure 8.1. Next, a prioritized and repetitive connection-oriented data service is supported; this corresponds to the media streaming sessions labeled MSS1 and MSS2 in Figure 8.1. The priority scheme permits otherwise standard data networking devices to obtain reserved media access at a rate of once per superframe repetition rate. *Superframes* are a sequence of two 10 ms frames. Through the use of priority, such applications as interactive videoconferencing and MP3 wireless headsets or remote speakers can be supported. Such connection-oriented data services commonly occur using UDP/IP flows.

The third service supported by the Home RF MAC layer is an isochronous, full-duplex, symmetric, two-way voice service. This service is used to map multiple voice connections that correspond to the DECT protocol.

Frame Operations

Because the Home RF standard can support up to eight simultaneous prioritized streaming media sessions and up to eight cordless voice connections, the composition of frames can vary considerably. In addition, when there are no prioritized streaming media or cordless voice connections, the MAC frame is limited to transporting asynchronous traffic. To obtain an appreciation of the manner in which the Home RF MAC frame can vary, let us turn our attention to examining several operational scenarios.

ASYNCHRONOUS TRAFFIC SUPPORT. Figure 8.2 illustrates the manner by which a series of frames limited to transporting asynchronous traffic can be transmitted at one frequency during the 20-ms hop duration. When data is limited to asynchronous traffic, each hop frequency is used to transport as much traffic as possible during the 20-ms period.

Figure 8.2
Transporting multiple
frames is limited to
asynchronous traffic
during one hop
duration.

Figure 8.2
Transporting multiple frames is limited to asynchronous traffic during one hop duration.

ADDING VOICE TRANSPORT. When isochronous traffic is added to support a mixture of traffic types, the duration of the frame length is reduced to 10 ms. Figure 8.3 illustrates the addition of isochronous traffic to the formerly all-asynchronous traffic data flow previously illustrated in Figure 8.2. In examining Figure 8.3, note that the beacon (B) is added to support isochronous traffic while the frame length is reduced to 10 ms. Also note that the isochronous traffic represents the transport of one mobile telephone call. Under the DECT specification, calls are transported using 32-Kbps *adaptive pulse code modulation* (ADPCM). According to the Home RF Working Group, the 32-Kbps ADCPM digital voice compression scheme results in a *Mean Optimum Score* (MOS) of 4.1. MOS represents a subjective rating system of voice quality that uses a scale of 1 to 5, with 5 representing the highest quality of voice. The reason MOS is a subjective rating is because of the manner in which an MOS rating occurs. Typically, people are placed in a relatively soundproof area and asked to listen to a prerecorded conversation spoken via a transmission method that employs the voice compression and reconstruction algorithm to be tested. Because every person has a different level of hearing acuity, the results are subjective. However, if a large base of people is used, the accuracy of the results becomes more meaningful. Returning to the 4.1 MOS rating, this is very close to toll-quality PCM obtained via a wired PSTN connection. Thus, the use of 32-Kbps ADPCM provides a relatively high quality of reconstructed voice.

In addition to supporting near-toll-quality voice, DECT provides support for such features as call line ID, call waiting, call forwarding, and other features associated with the wired PSTN. Up to four simultaneous active handsets out of eight active handsets can be supported under the Home RF standard.

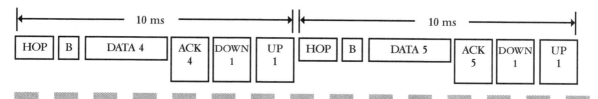

Figure 8.3   Adding isochronous traffic reduces the frame length to 10 ms.

Returning our attention to Figure 8.3, note that the two frames shown represent a Home RF superframe. Also note that the first frame is transmitted at one frequency while the second frame is transmitted at a different hop frequency.

To illustrate the effect of adding another voice call, consider Figure 8.4. In this example, we assume that asynchronous traffic continues to be transported while a second call is added. Note that as a call is added, the proportion of slot time available for transmitting data correspondingly decreases. Once one or both voice calls end, subsequent frames are automatically adjusted. For example, if one of the two voice calls terminates, subsequent frames continue to be 10 ms in length because one voice call continues to be active. If both voice calls end, subsequent frames expand to a duration of 20 ms, similar to the frame shown in Figure 8.2.

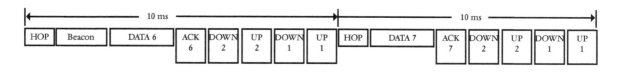

Figure 8.4   When a call is added, the 10-ms frame is automatically adjusted.

The Future

In this concluding chapter we uncover our crystal ball in an attempt to predict the direction the technology covered by this book will take. Our predictions will be more accurate than a fortune teller's crystal ball because we know the direction of work on additional standards and proprietary products that is currently being performed.

In this chapter, we briefly focus on several areas that can be expected to extend the capability of wireless LANs or even make the selection process of a particular LAN a bit easier. In doing so we discuss an FCC ruling and then turn to a few IEEE standards being developed; we conclude this chapter by examining a diagnostic testing program.

FCC Part 15 Ruling

During the early part of 2000, the Home RF Working Group proposed an extension of the maximum per-hop channel bandwidth from 1 to 3 and 5 MHz. These wider bandwidths were required to enable the frequency hopping spread spectrum (FHSS) method of communications employed by Home RF equipment to be developed to support higher operating rates.

Overview

During the latter part of 2000, the Federal Communications Commission (FCC) amended Part 15 of its rules for frequency hopping in the 2.5-GHz band. The amended Part 15 rule, as well as the original Part 15 rule, are compared in Table 9.1 In examining Table 9.1, note that the key difference between the original FCC Part 15 and amended FCC Part 15 rules are in the area of the number of available channels, maximum 20-dB bandwidth, and maximum transmit power level. Although the amended Part 15 rule only applies to FHSS in the unlicensed 2.4-GHz band, it has caused a degree of controversy even before Home RF products complying with the new regulations have been developed. The reason for the controversy is concern about potential interference between different products that operate in the 2.4-GHz band.

TABLE 9.1

FCC Part 15 Limits for FHSS in the 2.4-GHz Band

Parameter	Old Part 15	Amended Part 15
Frequency Bands	2.4000—2.4835 GHz	2.4000—2.4835 GHz
Number of Available Channels	75 nonoverlapping Channels	Minimum of 15 Nonoverlapping channels over a minimum span of 75 MHz
Maximum 20 dB Bandwidth	1 MHz	1, 3, 5 MHz
Maximum Transmit Power	1W	1 Watt 1 MHz 0.125 Watt 3 MHz 0.125 Watt at 5 MHz
Minimum Hopping Rate	2.5 Hops/s	2.5 Hops/s
Average Channel Dwell Time	0.4s over a 30s period	0.4s over a 30s period

ISM Band Use

Currently three wireless technologies share the use of the 2.4-GHz ISM band. Those technologies include IEEE 802.11 and 802.11b products, Bluetooth, and the evolving Home RF product line. The IEEE 802.11 standard supports both FHSS and direct sequence spread spectrum (DSSS) operations in the 2.4000- to 2.4835-GHz band, with all 802.11 products operating at 1 and 2 Mbps. In comparison, IEEE 802.11b products use DSSS at data rates of 5.5 and 11 Mbps and can fall back to operate at 2 and 1 Mbps, also using DSSS.

Bluetooth represents an on-the-go short range communications technology designed for incorporation into PDAs, cellular telephones, kitchen appliances, fax machines, and computers supporting symmetric data transmission at 432.6 Kbps and asymmetric data rates of 721 and 57.6 Kbps. Bluetooth also operates in the 2.4-GHz ISM band using FHSS. There are three power levels associated with a Bluetooth device—a Class 1 power level of 100 mW, and Class 2 and Class 3, which operate at maximum transmit power levels of 2.5 mW and 1 mW, respectively. Both Class 2 and Class 3 power levels represent "listening" levels of operation that conserves power.

The third wireless technology that operates in the 2.4-Ghz band is the Home RF Shared Wireless Access Protocol (SWAP). The Home RF SWAP standard specifies the use of FHSS with a maximum transmit power of 100 mW to provide a data rate up to 10 Mbps.

RF Interference

In addition to two 802.11 standards—Bluetooth and the Home RF SWAP standard occupying the 2.4-GHz band—we must consider the potential effect of other wireless devices. If you visit a Kmart or Wal-Mart and examine their home appliance section, you'll notice stickers on portable phones that inform you they operate in the 2.4-GHz band. In addition, if you are very inquisitive and examine the metal plate affixed on the back of many microwave ovens you might note that they also operate in the 2.4-GHz band. Thus, to put it mildly, the 2.4-GHz band is getting rather crowded, and the potential for RF interference between systems is increasing.

Before you have a panic attack regarding potential interference, it is important to note that many factors govern the potential degree of interference. In addition to operating frequency and hopping rate, the power level, operating range, signal bandwidth, and protocol must be considered. The widening of the Home RF bandwidth to 3 and 5 MHz results in an increased probability of a wideband signal's colliding with a narrowband signal. Because of this, a site survey of wireless technology within a geographic area surrounding a planned wireless LAN area of coverage might be in order. This is especially true if your organization rents part of a floor in a shared tenant building. In this situation, coordination with the building manager concerning the use of wireless technology could go a long way towards alleviating problems prior to their occurrence.

The power level of Bluetooth devices is minimal and their actual data exchange only occurs for short periods of time, so you can more than likely discount any potential interference between Bluetooth and 802.11 LANs as being very intermittent at most. However, because the Home RF standard will use FHSS with 3- and 5-MHz bandwidth, it is possible that products designed for compliance with the amended FCC Part 15 specification can cause more interference than Bluetooth devices. Thus, to control potential interference, you may wish to discuss the locations and use of other technology with your building manager or other departments within your organization. Similarly, because microwave ovens also operate in the 2.4-GHz bands, it is possible to note and easily avoid potential RF interference by considering the location of lunchrooms, break rooms, and other locations where microwave ovens might be found when considering the installation of a wireless LAN. Sometimes a thin lead shield around the back of a microwave oven is more than sufficient to block potential interference to a wireless LAN client installed in an office on the other side of a break room wall. Thus, a site review in preparation for the installation of a wireless LAN can resolve many problems before they materialize and eliminate interference before it becomes a problem.

The IEEE 802.11g Standard

The IEEE 802.11a standard extends the wireless LAN transmission rate to 54 Mbps. Unfortunately, in doing so it uses a completely different modulation scheme from the original 802.11 and the 802.11b standard extension.

Under the IEEE 802.11a standard, orthogonal frequency division multiplexing (OFDM) is specified, whereas 802.11b supports DSSS, which is also one of three access methods supported by the original 802.11 standard, the other two being FHSS and infrared. Another key difference in IEEE standards concerns the frequency band used. Both the IEEE 802.11 and 802.11b standards operate in the 2.4-GHz band. In comparison, the IEEE 802.11a standard specifies equipment operating in the 5-GHz frequency band.

Backward Compatibility Issues

The difference between the IEEE 802.11a standard and prior IEEE standards makes it difficult to provide backward product compatibility. To do so, an 802.11a device must support both OFDM and DSSS in two different frequency bands, which would increase the cost of an all-encompassing chip set. Perhaps recognizing this compatibility problem, the IEEE formed a task group to extend the operating rate of the 802.11b standard to 22 Mbps. The new standard extension will be known as the 802.11g standard.

The 802.11g standard extension is similar to the basic 802.11 standard and the 802.11b standard extension because it is also designed to operate in the 2.4-GHz band. Because the 802.11g standard extension is being developed to provide compatibility with 11 Mbps networks that use DSSS, I believes this extension will be more popular than the 802.11a extension. The first series of 802.11a and 802.11g products are expected to arrive during 2002. Thus, it may be a while before my prediction of popularity is proved one way or another. From the perspective of compatibility and expected cost, however, it is fairly easy to make a case for equipment that supports the 802.11g standard extension. Concerning compatibility, since the 802.11g standard will operate in the 2.4-GHz band, it is relatively easy to provide DSSS compatibility with 11 Mbps operations. But because the 802.11a standard operates in different frequency band, to obtain backward compatibility a second chip set would be required; adding to the cost and complexity of 802.11a-compatible devices.

Area of Coverage Consideration

A third reason why I believe that 802.11g equipment will have a greater acceptance than 802.11a devices concerns range. The range of a 5.4-GHz device is less than that of a 2.4-GHz device, output power being equal, because higher frequencies alternate more rapidly than lower frequencies. Because 5.4-GHz devices have a shorter range than 2.4-GHz products, this also means that an office would require more access points when operating an 802.11a network than with an 802.11g network.

As a general rule of thumb, you can expect an 802.11a access point to cover a radius of 90 feet, whereas an 802.11g access point will provide a radius of coverage of 200 feet or more. Because the area of a circle is πr^2, an IEEE 802.11a network requires approximately four times the number of access points a 802.11g network needs. Even if the cost of each access point were equivalent, an organization requiring multiple areas of wireless coverage could face a stiff bill for access points if it elected to use 802.11a products instead of 802.11g devices.

The IEEE 802.1x Standard

Because the IEEE 802.1x standard covers authentication, we must first review the limitations of the Wired Equivalency Protocol (WEP).

Earlier we noted that security on a wireless LAN occurs via the Wired Equivalency Protocol (WEP). WEP has several limitations, including the use of a 40-bit or 104-bit key that must be entered into each device. This key is then used in conjunction with a vector to produce a 64-bit or 128-bit key. Only the 64-bit key is required to be supported, and by default WEP is disabled on IEEE 802.11 LANs.

Another major limitation associated with WEP is the fact that secure communications between two devices requires a person to physically enter the key for each device. While this can be easily accomplished within most organizations, what happens if you are traveling with your 802.11-compatible notebook and check into a Hyatt Hotel that operates an Internet portal? If the hotel hands clients a note that says that today the WEP hex key is ABADABAD∅7, then every guest requiring wireless connectivity would know the WEP key to use. Not only does this action compromise the WEP key but it illustrates a problem in using a key for authentication. It is relatively easy for a person to determine the WEP key used by a public portal and to use that key with monitoring equipment to read other traffic and impersonate a legitimate user given enough time. Recognizing the need for a common method of authentication for both wired and wireless LANs resulted in the development of the IEEE 802.1x draft standard.

Overview

The IEEE 802.1x standard provides network-based authentication on a client accessing a LAN. Both wired LAN switches and wireless access points are supported by the 802.1x standard, whose official title is Port Based Network Access Control. The 802.1x standard, like many other standards, avoids the not invented here (NIH) syndrome by borrowing from prior standard efforts. The 802.1x standard specifies the manner by which the Extensible Authentication Protocol (EAP), which is specified in RFC 2284, should be encapsulated in LAN frames. EAP represents a general protocol that supports multiple authentication methods. Thus, its specification within the IEEE 802.1x standard permits prior effort to be used.

Operation

As a mechanism to provide authentication to obtain access to IEEE 802 LANs, the 802.1x standard ideally operates at the first point of attachment to a network. This means that access points

in a wireless environment as well as switch ports in a wired environment represent logical points where the 802.1x standard will
be supported.

Figure 9.1 illustrates the general operation of the IEEE 802.1x
standard in a wireless environment. Under this standard the supplicant represents the entity being authenticated and requiring
access to the services of the authenticator. A client (supplicant) initiates a conversation by requesting a connection through either
an Ethernet access point or a wired Ethernet port. This conversation occurs via EAP over LANs (EAPOL) to the authenticator,
which represents software residing on the switch or access point.
The authenticator requests the identity of the supplicant, who
then responds. The authenticator then accesses an authentication
server and informs the server that the supplicant is requesting
access. The server, which can be a Radius server, smart card server,
or another type of authentication server asks for proof of identify from the supplicant requesting access. If the proof of identify
is acceptable to the server, it then informs the switch or access
point to grant the supplicant access to the network. Otherwise
the supplicant is denied access to the network.

Figure 9.1
General operation of
IEEE 802.1x network
authentication
in a wireless
environment.

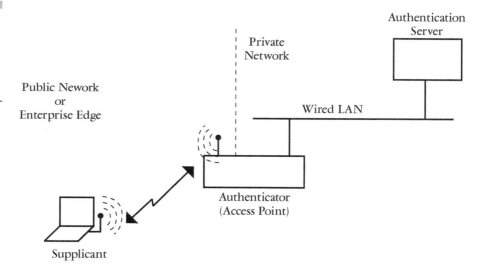

Although the authentication server can be collocated with the authenticator, more often than not it will probably represent an external server. That server can be a Radius server, Kerberos server, or a server that supports the use of smart cards, public key encryption, or even the use of one-time passwords.

Great Expectations

Although the IEEE 802.1x standard was not finalized when this book was written, I close my coverage of the standard under the heading "great expectations." The reason for the upbeat nature of this heading resides in the fact that although not finalized, the standard has gained the support of many movers and shakers in the field of data communications. In addition, Microsoft added support for 802.1x in its new Windows XP operating system. With support of the 802.1x standard gaining momentum, it can be expected to eventually represent the preferred method of authenticating wireless clients.

AiroPeek, A Wireless Protocol Analyzer

In this concluding section of this concluding chapter, we focus our attention on AiroPeek, a wireless protocol analyzer program from WildPackets, Inc., of Walnut Creek, California. AiroPeek represents an addition to the product line of a company many readers might remember as the AG Group. The AG Group developed the Ether-Peek software protocol analyzer that made its way into numerous organizations because of its easy use and comprehensive protocol decoding and statistical reporting capability. The AG Group changed its name to WildPackets during 2000 and added AiroPeek to its product line. AiroPeek provides a comprehensive diagnostic capability for examining IEEE 802.11b networks.

Overview

The version of AiroPeek I used was limited to supporting the Cisco Systems 340 Series PCMCIA wireless LAN adapter. However, by the time this book is published, support for Symbol wireless and Cisco 350 series adapters and perhaps other hardware can be expected to be added.

Figure 9.2 illustrates the initial display of AiroPeek when it is first used after its installation process is completed. Note the gauge in the lower left corner of the screen. That gauge is one of a pair that displays network utilization and packet flow in real time. You can easily change the gauge display to a numeric digital display by selecting "Value" instead of "Gauge" from the tab at the bottom of the display. Because AiroPeek was limited to supporting a Cisco Systems 340 series wireless network adapter, the dialog box titled "Select Adapter" indicates that that adapter was installed in my computer.

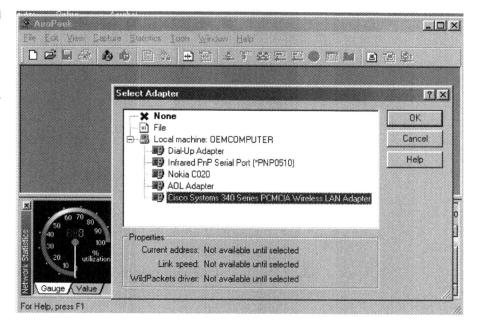

Figure 9.2
Selecting the hardware adapter for use by WildPackets' AiroPeek.

Capturing Traffic

The use of AiroPeek is similar to the use of any protocol analyzer in that it requires traffic to analyze. To obtain traffic for analysis you can either import a file of previously captured traffic or commence packet capturing. In the latter situation, you use the Capture menu, resulting in the display of the dialog box illustrated in Figure 9.3.

In examining Figure 9.3, note that there are several packet capture options to consider. You can invoke a continuous capture of over-the-air packets, limit the amount of data captured for each packet read, and reset a buffer size whose initial value is 4 Mb. Because packet capture is limited to 64 bytes by default, the user can concentrate monitoring on the header and record more packets.

Figure 9.3
Packet capturing requires several items to be considered.

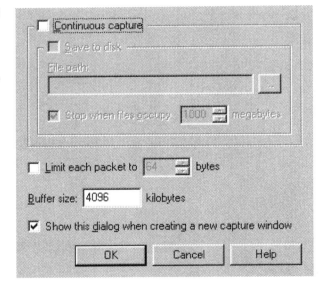

Since the proof of the pudding is in the eating, let us try capturing some off-the-air traffic. Figure 9.4 illustrates the AiroPeek capture window at a point when 432 packets have been captured. In examining Figure 9.4, note that at the time the screen was captured, packet information was being displayed. If you look above the gauges, you will see the tab labeled "Packets" is highlighted. By clicking on a different tab you can display information about nodes, protocols, conversations, and similar data.

If you turn your attention to the packet information shown in Figure 9.4, you note all broadcasts have the same source address. This is because the access point I used, as well as other IEEE 802.11b access points, periodically broadcast their basic service set ID (BSSID) as a mechanism to inform clients of their presence. The broadcasts occur at a 1-Mbps data rate while actual traffic transmitted occurred at 11 Mbps.

Figure 9.4
Viewing a summary of packets captured through the use of AiroPeek.

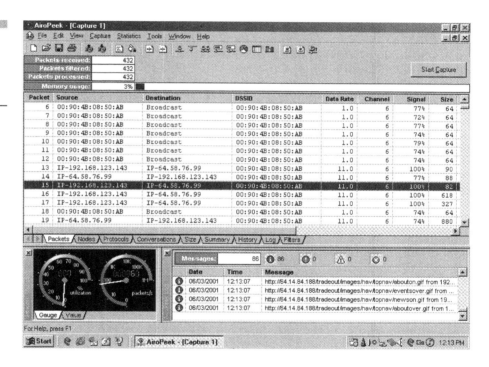

The IEEE 802.11b standard supports DSSS at data rates of 1, 2, 5.5, and 11 Mbps. Because the packet display also indicates the signal strength, it is possible to use AiroPeek as a mechanism to check client-access point connections in addition to its primary role as a packet analyzer.

If you look at packet 15, which is highlighted, you note its source IP address is 192.168.123.143. This is an RFC 1918 Class C address that represents a client behind a combined access point/broadband wireless router that was in the process of communicating with Yahoo!. I know the client was communicating with Yahoo! because the IP address of 64.58.76.99, which is shown as the destination, is the address that displays the Web page to sign onto Yahoo! mail. If you examine packets 16 through 19, you note that packets 16 and 17 continued sending information to Yahoo!, packet 18 represents a periodic access point broadcast, and packet 19 represents the reply from Yahoo!.

Protocol Summary

Tailoring a network for optimum performance requires knowledge about the traffic being carried. For example, a large number of DNS operations may indicate that a local DNS server might be a better option than depending on the use of an ISP's server. To determine the type of traffic flowing on a network, wireless or wired, requires the ability to note the protocols being used. Much like its wired cousin EtherPeek, AiroPeek includes a protocol summary capability that permits a user to break out by percentage different types of traffic. Figure 9.5 illustrates the display of the screen after the tab labeled "Protocols" was selected. Note that at this point 2643 packets were captured, which is also shown as packets filtered and processed. AiroPeek includes a comprehensive packet filtering capability that enables the user to specify one or more metrics to be used for packet capturing. For example, you could filter based on a protocol, source address, destination address, or even a combination of metrics.

One of the key uses of packet filtering is the ability to focus your attention on a specific metric of interest instead of having to sort through a large number of captured packets. By carefully considering your monitoring requirements, you can create an applicable filter that will facilitate the examination of packets that are appropriate to the problem at hand.

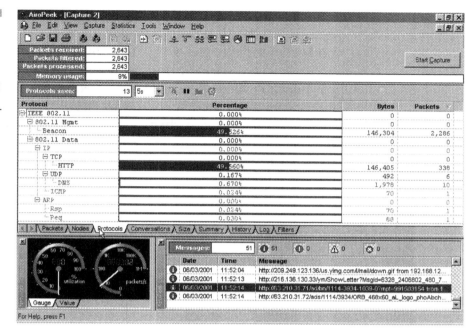

Packet Decoding

In concluding our examination of AiroPeek, we take a brief look at its packet-decoding capability. AiroPeek maintains many of the capabilities of EtherPeek, including the ability to double click on a captured packet to decode the contents of the selected packet. To be faithful to terminology as well as the air environment, AiroPeek captures and provides a decoding capability that operates on frames. However, wireless frames transport packets, so I describe the decoding of packets as a feature of the program. AiroPeek provides a decoding capability that allows users to observe the layer 2 header as well as the headers of the higher-layer protocols being transported within the frames that were monitored.

Figure 9.6 illustrates a portion of the packet decoding capability of AiroPeek. In this example, a previously captured packet was selected for decoding and I positioned the decoding window so that the IP header was at the top. Prior to display of the IP header information, the window shows the layer 2 MAC decode. In examining the IP header, decode you note that AiroPeek provides both the decode and a hex window that corresponds via highlighting to the selected portion of the decode window. For example, because the IP header was selected in the decode window, the hex values associated with the IP header are highlighted in the hex window. You can check this by noting that the version and header length fields whose values were decoded as 4 and 5 respectively correspond to the values 45 as the first portion of the highlighted entries in the hex window.

Figure 9.6
Using AiroPeek to decode the IP header of a packet.

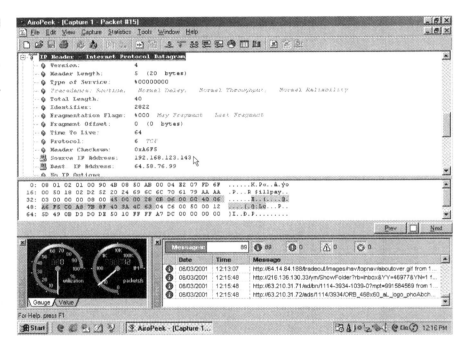

When strange events occur, your first line of knowledge for resolving the situation is the ability to decode the contents of packets flying through the air, a situation where AiroPeek excels. For organizations that must stay on top of the situation and require a diagnostic analysis capability as well as the ability to note statistics and usage of their wireless LAN, a wireless protocol analyzer is a necessity.

APPENDIX A

HARDWARE MANUFACTURERS

This appendix provides you with a single point of reference concerning manufacturers of wireless LAN products. Vendors are listed in alphabetical order. Although I explored the Web site of each vendor to extract relevant contact information, you should note that the most important item listed for each site is probably the Web address. By searching the Web site of each vendor, you may be able to obtain information concerning the latest products offered by each vendor. In addition, several vendors offer online shopping so that you can purchase a product or series of products directly. Furthermore, if you already have one or more products, you may wish to consider periodically visiting the appropriate vendor Web site to obtain software updates and new drivers.

In addition to purchasing wireless products through conventional sales channels and online vendor stores you have another option, online auctions. During the preparation of this book I periodically checked the eBay Web site (www.ebay.com) and noted that many wireless LAN adapter cards and access points were sold for approximately half their suggested retail price. Because I discovered several 'tricks' associated with locating wireless LAN products on eBay, at the conclusion of this appendix a separate section discusses how to locate applicable equipment on this online auction Web site.

AGERE SYSTEMS
(formerly part of Lucent Technologies)
555 Union Blvd.
Allentown, PA 18109
Telephone: 1-800-372-2447
1-610-712-5130
Web site: www.agere.com

ALVARION
(formed by the merger of Breezecom and Floware)
5858 Edison Place
Carlsbad, CA 92008
Telephone: 1-760-517-3100
Fax: 1-760-517-3200
Web site: www.alvarion.com
E Mail:n.america-sales@breezecom.com

CISCO SYSTEMS
170 West Tasman Drive
San Jose, CA 95134
Telephone: 1-408-526-4000
Toll free: 1-800-326-1941
Fax: 1-408-526-4100
Web site: www.cisco.com
Email: tac@cisco.com

D-LINK
53 Discovery Dr.
Irvine, CA 92618

Sales
Telephone: 1-800-326-1688
Fax: 1-949-753-7033
Web site: www.dlink.com
Email: sales@dlink.com

System
Telephone: 1-888-DLINK-SI (354-6574)

Fax: 1-949-753-7040
Email: si@dlink.com

Educational sales
Telephone: 1-800-326-1688x5329
Email: edusales@dlink.com

Technical support
Telephone: 1-949/790-5290
Email: ustech@dlink.com

INTEL CORPORATION
2200 Mission College Blvd.
Santa Clara, CA 95052
Telephone: 1-408-765-8080
Fax: 1-408-765-9904

LINKSYS
Linksys Headquarters
17401 Armstrong Ave.
Irvine, CA 92614
Telephone: 1-949-261-1288
Fax: 1-949-261-8868
1-800-546-5797 (LINKSYS)
Web site: www.linksys.com
Email: sales@linksys.com
Email Technical Support: support@linksys.com

NETGEAR INC.
4500 Great America Parkway
Santa Clara, CA 95054
Telephone: 1-408-907-8000
1-888-NETGEAR (1-888-638-4327)
1-800-211-2069
Fax: 1-408-907-8097
Email support: support@netgear.com

PROXIM
510 DeGuigne Drive
Sunnyvale, CA 94085
Telephone: 1-800-477-6946
1-408-731-2640
1-408-731-2700
1-800-229-1630
Web site: www.proxim.com
Email: sales@proxim.com

SMC NETWORKS
Web site: www.smc.com
Sales:
Email: sales@smc.com
Telephone: 1-800-SMC4-YOU
Technical support:
Email: techsupport@smc.com
Telephone: 1-800-SMC4-YOU

SOLECTEK
Solectek Corporation
6370 Nancy Ridge Dr., Suite 109
San Diego, CA 92121-3212
Telephone: 1-858-450-1220
Fax: 1-858-457-2681

SYMBOL TECHNOLOGIES
Symbol Technologies, Inc.
340 Fischer Ave.
Costa Mesa, CA 92626
Telephone: 1-800-722-6234
Web site: www.symbol.com
United States
Telephone: 1-714-549-6000
Service Telephone: 1-714-549-6000
Fax: 1-714-549-3431

Symbol Technologies, Inc.
Sales Office
141 Union Blvd., Suite 140
Lakewood, CO 80228
United States
Telephone: 1-303-989-6331
Fax: 1-303-989-6983

Symbol Technologies, Inc.
407 Wekiva Springs Rd. ,Suite 100
Longwood, FL 32779
United States
Telephone: 1-407-862-5000
Fax: 1-407-862-6126

3COM
Corporate Headquarters
5400 Bayfront Plaza
Santa Clara, CA 95052
Telephone: 1-408-326-5000
Fax: 1-408-326-5001
Product Sales Inquiries
1-800-NET-3Com
(1-800-638-3266)
Web site: www.3com.com

Locating Wireless Equipment on eBay

In concluding this appendix, I'd would like to share several techniques you can consider using to locate applicable wireless equipment. In addition, a bidding strategy that I periodically encountered is discussed because its utilization can make the difference between success and failure with respect to obtaining what you seek at an appropriate price.

Equipment Location

eBay reminds me of a giant Woolworth's of old. While Wool-worth's certainly stocked almost everything a person might need for the home, it was often difficult to locate an item of interest. Similarly, eBay may have what you need, but if the item is hidden from view, you may miss bidding on it.

To illustrate some equipment location techniques, let's take a tour of the use of the Search facility built into the eBay Web site. If you read this book, you noted that the IEEE 802.11 standard represents the most popular method of constructing wireless LANs. Well, if you use the search term 'IEEE 802.11' as shown in Figure A.1 in an attempt to locate wireless hardware, what level of response do you think you will achieve?

Although you might expect some items to be located through the use of this search term, in actuality the result was a shocking lack of matches! Figure A.2 shows the results of our search attempt and the response from eBay that "no items could be found for ieee 802.11."

Figure A.1
Using the search title IEEE 802.11 to locate hardware on eBay.

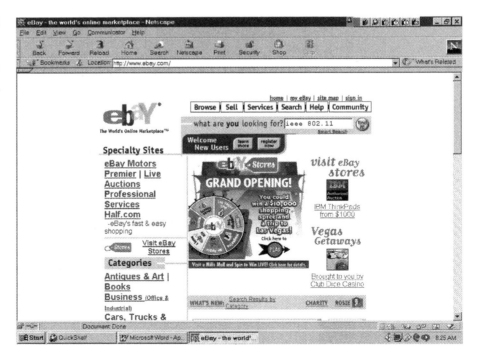

Figure A.2

The search response to our query for IEEE 802.11 returned no matches.

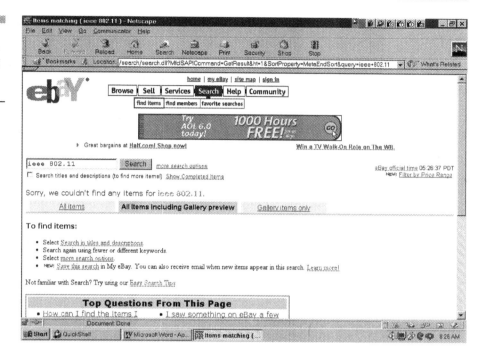

If you examine how items are listed on eBay, the results of my initial search are actually less than shocking. Rarely is a full descriptor of a standard used to describe an item. Let's go back to the drawing board and try some different queries to ascertain if eBay has wireless hardware for sale.

Because the search on "IEEE 802.11" did not return any items, let's drop the term "IEEE" and perform our search using the search term "802.11." Figure A.3 illustrates the use of the revised search term, which located ten items on eBay. For those new to eBay, the second, third, and fifth items listed have small icons to the right of the item description. Those icons as well as other icons not currently displayed indicate different things about a particular item. For example, the Lucent WaveLAN PCMCIA 802.11 adapter has the term "Buy Now" which means that instead of bidding on the item you can directly purchase it and in effect terminate the auction. For those interested, the Buy Now price was $67 while the high bid was $63 when the screen was captured.

Figure A.3
Revising an eBay
search using the term
"802.11."

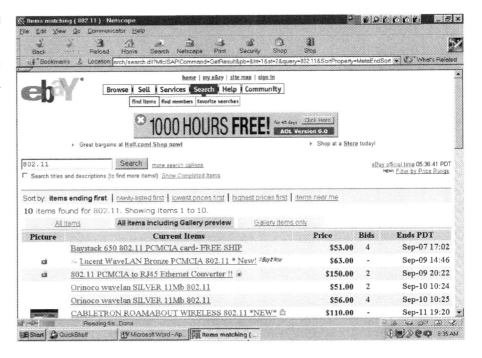

Figure A.3
Revising an eBay search using the term "802.11."

The icon on the next line indicates that the listing can be paid for using eBay Online Payments. eBay Online Payments enables eBay sellers to accept credit cards or Electronic Checks from winning bidders. Any winning bidder can pay safely online using Visa, Mastercard, or Discover credit cards and Electronic Check for an auction that includes eBay Online Payments as a payment method.

Continuing our short tour of icons, the next line in Figure A.3, which is for Cabletron hardware, includes an icon that represents a Gallery. The Galley is a new way of browsing items for sale at eBay. The Gallery presents miniature pictures, called *thumbnails,* for all of the items for which sellers have supplied pictures in JPG format. The Gallery displays the same information as the regular listing pages, but it lets you shop like a person browsing through a catalog, viewing items without having to click on their title.

Returning to our quest to locate wireless hardware, let's pretend we are from Missouri, the Show Me state. Let's not believe that there are only ten items associated with 802.11 wireless LANs and consider other methods to display applicable auction listings. Since LAN adapter cards are an integral part of wireless LANs, let's revise our search term and use "wireless LAN card" as a new search term.

Figure A.4 illustrates the use of the new search term "wireless LAN card" and the response to this search. As Superman would say, Holy Toledo, since we have now located 70 wireless LAN cards that are up for auction.

Figure A.4
Using the search term "wireless LAN card" to locate 70 auctions in progress.

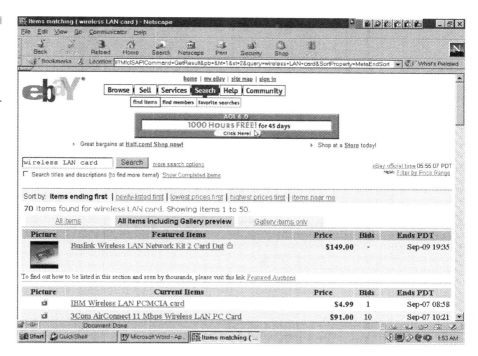

If you examine the last line shown in Figure A.4 you note the current bid for a 3Com AirConnect 11-Mbps wireless LAN card is shown as $91 and a total of ten bids had occurred on this particular auction when the screen was captured. Whether or not a bid of $91 represents a buy or a high price depends on several factors. First and foremost, read the listings about the equipment.

While some hardware can be found that is new, in the original box, with documentation, other hardware may represent used items that may or may not be operational. Another thing to consider is the price of the product if you purchased it directly. I suggest you surf the vendor sites noted earlier in this appendix to locate items you need and ascertain the cost to obtain the products directly from the manufacturer. You can then use such information to determine if you might be able to obtain a good deal or if the exuberance of auction bidding is making you irrational—sort of like Dr. Greenspan's irrational exuberance.

Although the use of the search term "wireless LAN card" was certainly an improvement over the other search terms, you might be curious why we just didn't type in the term "wireless" to locate everything. That search would indeed give us the listing for every wireless device auction on eBay, including wireless mice, keyboards, speakers, and so on. When I used the search term "wireless" I wound up with over 66 pages of auctions to sift through. So, unless you're curious about all wireless devices offered for sale, you more than likely want to narrow your search term. However, you also need to recognize that too narrow a term produces a limited number of items.

We probably want to consider the other major component of wireless LANs, the access point. Thus, we revise our search and enter "wireless access point" for the search title. The result of this search is shown in Figure A.5.

In examining Figure A.5, note we now located 85 auctions for wireless access points that are completely different from the auctions for wireless LAN cards. Thus, specificity does make a difference. In fact, if you entered search terms such as "wireless bridge" and "wireless antenna" you would obtain distinct listings for such equipment that would be different from previously located equipment. Now that we can locate items on eBay, let's talk about the bidding process.

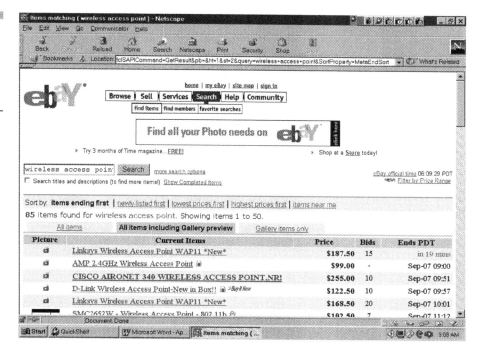

Figure A.5
Using the search term "wireless access point" resulted in the location of 85 auctions on eBay.

The Bidding Process

When you click on an item of interest on eBay you note the time remaining for the auction. If you scroll further down the page for an item, you come to the bidding section. Figure A.6 illustrates an example of that section for a D-Link Wireless Access Point, which was being sold as new. Some items on eBay have a *reserve price*, which means that unless the bid exceeds that price, you can be the highest bidder but not get what you want. If the product has a "buy it now" option, you can directly purchase the product and close out the auction. If you examine Figure A.6, you could purchase the D-Link Wireless Access Point for $175 or you could make a bid on it of at least $125, since the current bid is $122.50 and the bid increment is $2.50.

If you do not need the product immediately, you may wish to consider the following bidding strategy. First, determine the price you want to pay. Next, complete the screen, click on the button labeled "Review bid" and fill out the following screen a minute or less prior to the auction's closing. Then, about 30 seconds prior to the close of the auction submit your bid. While others may follow similar techniques, this action hides your intentions till the last moment and may make it possible to locate the bargain you need.

Figure A.6
By bidding on an item very close to the end of an auction you may increase your chance of becoming a successful bidder.

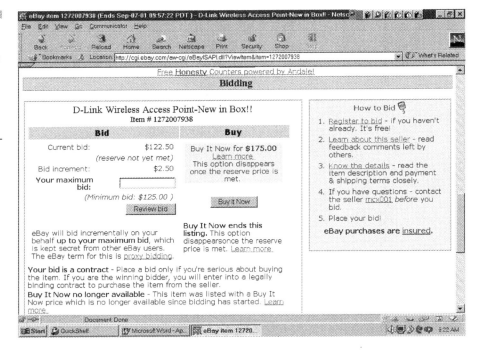

APPENDIX B

WIRELESS LAN ECONOMICS

This appendix provides readers with a frame of reference concerning the economics associated with constructing and operating a wireless LAN. In doing so we examine several scenarios, beginning with the economics associated with a very small wireless LAN and literally working our way upward to examine networks that support a large number of clients.

A word of caution is in order concerning the costs associated with different wireless LAN products mentioned in this appendix. Cost used in this appendix represents the average cost of hardware products and, in effect, can be considered to represent the cost of generic products. The average cost computed for wireless LAN products mentioned in this appendix was computed in late 2001.

Because the cost of most communications hardware products can be expected to decline over time, the costs associated with IEEE 802.11b products can also be expected to decline. The representative examples of wireless LAN networks presented in this appendix can also be expected to decline in price over time, making each network scenario more economical. However, because no IEEE 802.11a products were being marketed when this book was prepared, readers should note that the cost of new technology resembles a decaying pulse over time, with a relatively high unit cost when a product is introduced followed by a declining price as economy of production and competition perform as expected in a free-market economy.

Limited Client-Based Wireless LAN

The most elementary type of wireless LAN is established by installing a wireless LAN card in two individual PCs. This action results in the construction of a peer-to-peer wireless LAN, also referred to as an ad hoc wireless LAN.

When two PCs are connected in a peer-to-peer relationship, you actually obtain more than the ability to use network sharing under Windows to access files between computers. One of the little-used features supported by Microsoft Windows Internet sharing capability is *Internet Connection Sharing*.

Internet Connection Sharing enables you to configure your home or small-office computer network to share a single connection to the Internet. Using Internet Connection Sharing, one computer on your network, called the Connection Sharing computer, has an Internet connection and provides private IP addresses and name resolution services for one or more additional computers on the home or small-office network. Then, the other computers on your network can access the Internet through the Connection Sharing computer using private IP addressing translation.

When a computer on your network sends a request in the form of an IP datagram to the Internet, its private IP address is transmitted to the Connection Sharing computer, which translates it to the Internet IP address of the Connection Sharing computer, in effect performing a network address translation function. The Connection Sharing computer then transmits the datagram onto the Internet. When the results are returned, the Connection Sharing computer translates the IP address back again and routes the datagram to the correct computer on your network. Thus, the only computer on your home or small-office network that is visible to the Internet is the Connection Sharing computer. None of the other computers on your home or small office network has a direct connection to the Internet.

When automatic addressing is enabled, Internet Connection Sharing uses the Dynamic Host Configuration Protocol (DHCP) to dynamically assign private IP addresses to all computers on your home or small-office network. You can also disable the automatic addressing service and statically assign an IP address to each computer on the network.

Prior to using Microsoft's Internet Connection Sharing, all computers to be interconnected on your home or small-business network must have wireless network adapters installed. You need to select the computer that will be the Connection Sharing computer and use it to establish the connection to the Internet using your preferred method, such as DSL or a cable modem. Once you select the Connection Sharing computer, run the Internet Connection wizard to help you establish your connection. After your Internet connection is established and verified to be working, you install Internet Connection Sharing, then configure the Connection Sharing computer and all other computers on the network. Its also worth noting that you can configure computers on your network to use file and print sharing, so that they can access these resources from each other. Now that we have an appreciation for Internet Connection Sharing, let's focus our attention on the costs associated with this method.

At the time this book was prepared, the average cost of an 11-Mbps wireless PC card was $100 for a PC card format and $125 for a PCI bus-based network card. Thus, a two-client network consisting of one desktop computer and one notebook would cost $225. If you need to expand this type of network you would do so by adding either PCI bus-based network cards or PC cards, or both, depending on your hardware platforms. Because the Connection Sharing computer must perform a variety of communications functions in addition to any normal functions, the use of this method of wireless networking is probably best when you have only a handful of clients. If you have more than five clients or do not want to dedicate a $1000 PC to serve as an Internet Connection Sharing device, you should then consider the use of a wireless access point. If your wireless activity includes access to the Internet, you must consider the use of a wireless access point that has a built-in routing capability.

Access Point/ Router-Based Wireless LAN

A second type of wireless LAN is based on the use of a combined access point/router. This type of communications hardware typically provides one Ethernet port or a USB port that furnishes a connection to a cable or DSL modem. The access point/router may also include a limited number of 10/100-Mbps Ethernet switch ports.

The use of an access point/router should be considered when peer-to-peer networking will not satisfy your communications requirements. In addition, the use of an access point/router is normally less expensive than dedicating a personal computer as an Internet sharing device. In a home environment, the use of an access point/router should be considered if you do not wish to dedicate a PC to be always on to provide Internet access. In an office environment, you should consider the use of an access point/router if your organization has more that four or five computers that require simultaneous access to the Internet.

The average cost of a combined access point/router varies based on device functionality. A basic access point/router without any Ethernet switch ports has an average cost of $225. In comparison, the average cost of a combined access point/router with three built-in 10/100-Mbps Ethernet ports is $300. Using the average cost of a wireless PC card as $100 and the average cost of a PCI wireless LAN card as $125, it is easy to determine the potential outlay to establish a wireless LAN based on the use of a single combined access point/router. For example, if you anticipate ten notebook-based clients, eight desktop-based clients, and one combined access point/router without Ethernet switch ports, your total expected cost would become :

Access point/router	$225.00
10 PC Card @ $100	$1000.00
8 PCI wireless cards @125	$1000.00

Your total expected cost to establish an IEEE 802.11b wireless LAN to support 18 clients accessing the Internet would be $2225.

Wired LAN Access

There are two methods you can consider to provide wireless access to a wired LAN. If you need access to both an existing wired LAN and access to the Internet you should consider the use of a combined access point/router that includes Ethernet switch ports. Then, you could cable one or more switch ports from the combined access point/router to your wired LAN. If your organization currently obtains access to the Internet via a wired LAN you could use an access point and cable the access point to your organization's wired LAN.

When examining different types of access points while preparing this book, I noted a wide range of costs. Some access points that did not include space diversity antennas could be obtained for approximately $175. Other access points that included space diversity antennas as well as support for 128-bit wireless equivalency privacy (WEP) security had a retail price of $700.

You must consider the number and type of wireless client devices and then compute the cost for providing such devices with a wireless communications capability. Once this is accomplished, you must consider the cost of the combined access point/router or access point required to support your clients.

To illustrate the procedure for determining the cost required to support the connection of wireless clients to an existing wired LAN let's assume we have twenty clients, twelve representing desktop computers and eight representing notebook computers. Let's also assume our organization requires 128-bit WEP and will obtain this capability through the use of a $700 access point. Thus, the expected cost is:

12 wireless LAN PCI Cards for desktops @$125	$1,500
8 wireless LAN PC Cards for notebooks @$100	$800
1 access point supporting 128-bit WEP	$700
Total estimated cost	$3,000

APPENDIX C

PRACTICAL COMMUNICATIONS SECURITY

This appendix provides you with a practical technique to add a degree of communications security to your wireless LAN.

If you read the *New York Times* or *The Wall Street Journal* during 2001 concerning wireless LANs, you probably noted stories of how two men in a van were able to travel from parking lot to parking lot in Silicon Valley and, using readily available equipment, capture wireless communications. According to those articles, the two men in a van were able to read all traffic without any need to turn to high mathematics to decode encrypted communications because, by default, the Wired Equivalency Protocol (WEP) is disabled. Thus, as previously noted in this book, at a minimum, if you want a degree of security you need to enable WEP.

According to more recent trade journal articles, several scientists and mathematicians announced that WEP has several flaws that provide unscrupulous people with the ability to rapidly decrypt encrypted transmission. While I cannot comment on those trade journal articles because they were summaries rather than detailed procedural information, it is obvious that if you cannot read a signal you cannot determine what is being communicated. Thus, the purpose of this appendix is to provide you with a simple technique that may provide a degree of transmission intercept immunity that will, in turn, make life much harder for those who wish to read your wireless LAN communications.

If you examine the configuration of a wireless LAN access point or a combined wireless LAN access point and router, you will note the device has either single or dual antennas. Most equipment vendors tell you to mount the access point or combined access point/router in a central location to provide an optimal level of signal strength to all clients. Although this method of equipment location is correct with respect to signal strength, it does not consider how radiation can leak out of a building into a parking lot. But there is an easy-to-use technique to minimize the leakage of radiation.

To minimize radiation leakage outside your building, position your access point or combined access point/router at a location where signal strength is minimized to the outside. But because this positioning could adversely effect the signal strength flowing to reliable clients we need a better method. That method, believe it or not, uses a roll of tin foil. As an example, I located my access point in my kitchen, which is next to my garage. After shielding the antennas of the access point with a doubled sheet of tin foil, I went into my garage with a second laptop to determine the level of signal strength that could be monitored. Using the Wireless LAN Configuration Utility Link Info tab, which is shown in Figure C.1, I had a hard time picking up anything beyond a very weak access point signal. As I moved out of my garage onto my driveway, the signal disappeared.

While this shielding still leaves the signal strength of client transmissions unaffected, with a bit more use of tin foil it is possible to shield an area where communications leakage flows into a public area, such as a parking lot. Thus, a roll of tin foil or a similar shielding material, at a cost of a few dollars, may be all that is necessary to place a limit on communications leakage.

Figure C.1

Using the SMC Networks Wireless LAN Configuration Utility Link Info tab to determine signal strength.

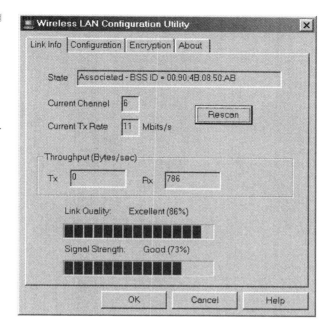

GLOSSARY OF TERMS
AND ABBREVIATIONS

802.1x A standard that defines how network-based authentication is provided for clients accessing wireless or wired LANs.

802.11 The fist IEEE wireless LAN standard which defines operating rates of 1 and 2 Mbps for frequency hopping spread spectrum, direct sequence spread spectrum and infrared operations.

802.11a An extension to the IEEE 802.11 standard that defines operations at data rates up to 54 Mbps in the 5-GHz frequency band.

802.11b An extension to the IEEE 802.11 standard that defines operations at data rates of 5.5 and 11 Mbps using direct sequence spread spectrum as well as downward compatibility with the 802.11 standard using DSSS at 1 and 2 Mbps.

802.11g An extension to the IEEE 802.11 standard that extends the operating rate of 802.11b devices to 22 Mbps.

access point A device that connects wireless devices to a wired infrastructure. To perform this operation the access point functions as a bridge.

ACK Acknowledgement.

Ad hoc LAN A group of two or more wireless LAN devices that communicate directly with each other and not through an access point.

AM Amplitude Modulation.

Amplitude modulation (AM) The process of altering the magnitude of a signal to represent binary 0s and binary 1s.

Antenna diversity The use of two antennas separated by an odd multiple of a quarter wavelength to receive the highest level of signal power while ignoring multipath reflections.

Attenuation A decrease in the waveform of a signal as it propagates from an antenna.

Authentication The process of verifying the identity of a client.

bandwidth A range of continuous frequencies. If f1 is the highest frequency in a band and f2 the lowest, then the bandwidth is f1-f2.

Barker sequence An 11-bit spreading code used under the IEEE 802.11 standard for Direct Sequence Spread Spectrum communcations.

Basic Service Set (BSS) The communications domain for an access point under the IEEE 802.11 standard.

Basic Service Area (BSA) The area of wireless coverage within which members of a Basic Service Set (BSS) can communicate.

Baud The rate of signaling change commonly expressed in Hertz.

Beacon A frame periodically transmitted by an access point which identifies its presence to stations.

Bel The ratio of power transmitted to power received such that $B = \log_{10} P_0/P_I$, where P_0 represent power output or received while P_I represent input or transmitted power.

BER Bit Error Rate.

Binary Phase Shift Keying (BPSK) A modulation technique under which the phase is altered between two values to correspond to binary 0s and binary 1s.

Blowfish A symmetric block cipher encryption algorithm developed by Bruce Schneier of Counterpane Internet Security. Blowfish supports a variable length key from 32 bits to 448 bits.

Bps   Bits per second.

BPSK   Binary phase shift keying.

BSA   Basic Service Area.

BSS   Basic Service Set.

Carrier   An oscillating radio frequency signal which is modulated to convey information.

CCK   Complementary Code Keying.

Chip   The mapping of data bits into a spreading code pattern.

Chipping code   The sequence of bits used to spread a narrowband signal into a wideband signal under direct sequence spread spectrum communications.

Complementary code   A pair of equal length sequences such that the number of pairs of like elements with any separation in one series is equal to the number of pairs of unlike elements with the same separation in the other series.

Complementary Code Keying (CCK)   A set of 64 8-bit codewords that's used as a spreading function.

Connection point (CP)   A Home RF gateway that provides access from a wireless network to a wired infrastructure as well as the public switched telephone network.

CP   Connection point.

CPS   Cycles per second.

CRC   Cyclic Redundancy Check.

CTS   Clear to Send.

DAB   Digital Audio Broadcast.

dB   Decibel.

dBm   Decibel above one milliwatt.

DBPSK   Differential Binary Phase Shift Keying.

DCF   Distributed Coordination Function.

DECT   Digital Enhanced Cordless Telephone.

Digital Enhanced Cordless Telephone (DECT)   The cordless telephone standard.

DQPSK   Differential Quadrature Phase Shift Keying.

Decibel (dB)   A more precise measurement of power gain or loss than the Bel. The power measurement in decibels (dB) defines the ratio of power transmitted to power received such that $B = 10 \times \log_{10} P_0/P_I$, where P_0 represent power output or received while P_I represent input or transmitted power.

decibel above one milliwatt (dBm)   The ration of output power to a 1 mw input power test tone.

Demodulation   The reverse process to modulation, which converts an analog signal back into digital form.

DHCP   Dynamic Host Configuration Protocol.

Dibit encoding   The process of encoding two bits into one signal change.

Differential Binary Phase Shift Keying (DBPSK)   A modulation technique under which each data bit is mapped into one of two phase changes.

Differential Phase Shift Keying (DQPSK)   A modulation technique in which pairs of bits are mapped into one of four phase changes.

DIFS   Distributed Inter Frame Space.

Direct sequence spread spectrum (DSSS)   A transmission technique under which a narrowband signal is spread over a wider frequency band.

Discrete multitone (DMT)   Aa transmission method under which a series of carriers at right angles to one another are modulated across a frequency band. This represents another term for orthogonal frequency division multiplexing.

Distributed Coordination Function (DCF)   The official name of the access method used by IEEE 802.11 wireless LANs.

Distributed Inter Frame Space (DIFS)   A period of time that a wireless station waits prior to being able to transmit.

Distribution system   The infrastructure that connects Basic Service Sets together to form an Extended Service Set (ESS).

DMT   Discrete multitone.

DSSS   Direct Sequence Spread Spectrum.

Dwell time   The length of time a signal is transmitted at a particular frequency under Frequency Hopping Spread Spectrum communications.

Dynamic Host Configuration Protocol (DHCP)   A protocol that leases IP addresses to stations.

EAP   Extensible Authentication Protocol.

EAPOL   Extensible Authentication Protocol over LANs.

EHF   Extremely High Frequency.

EIFS   Extended Interframe Space.

ELF   Extremely Low Frequency.

EOF   Elecro Optical Frequency.

ESS   Extended Service Set.

ESSID   Extended Service Set Identification.

ETSI   European Telecommunications Standards Institute.

Extended Interframe Space (EIFS)   A time delay for transmission used by a station that received a frame it could not understand.

Extended Service Set (ESS)   The configuration of two or more Basic Service Sets to function as an entity to permit roaming.

Extension point   A special type of repeater used to extend the transmission distance of a wireless LAN.

Fading   An obstruction to a signal that weakens its reception at a receiver.

Fast Ethernet   A version of Ethernet that operates at 100 Mbps.

Fast hop   A Frequency Hopping Spread Spectrum communications system in which the hopping rate exceeds the frame rate.

FCC   Federal Communication Commission.

FEC   Forward Error Correction.

FHSS   Frequency Hopping Spread Spectrum.

FM   Frequency Modulation.

Fps   Frames per second.

Frames per second (FPS)   The rate at which a transmission system places frames onto the air.

Frequency   The number of periodic oscillations or waves that occur per unit time.

Frequency Hopping Spread Spectrum (FHSS)   A transmission technique under which a transmitter hops from one frequency to another, transmitting on each frequency for a predefined duration.

Frequency modulation (FM)   The process of varying the frequency of a carrier signal in tandem with changes in the composition of digital data.

Frequency Shift Keying (FSK)   The process of modulating data by generating one of two tones or frequencies, shifting tones in correspondance to the composition of each bit.

Frequency spectrum   The range of frequencies up to 10^{23} Hertz.

FSK   Frequency shift keying.

Gateway   A term used in TCP/IP configuration screens to reference a local router that serves a workstation.

Gaussian Frequency Shift Keying (GFSK)   A version of FSK modulation under which broadband pulses are passed through a Gaussian filter prior to modulation occurring.

Gaussian noise   Another term for thermal noise.

Giga   Billion.

GFSK   Gaussian Frequency Shift Keying.

GHZ   Gigahertz.

Gray code   A binary code which has the property that, between any two successive binary numbers, only one bit changes state.

HEF   High Energy Frequency.

Hertz   A term used to represent one cycle per second in honor of the German physicist.

HF   High Frequency.

Hidden node   A station in a service set that cannot detect the transmission of another station and thus fails to recognize that the media is busy.

ICMP   Internet Control Message Protocol.

IEEE   Institute of Electrical and Electronic Engineers.

Impulse noise   Noise resulting from such periodic disturbances as lightning, solar flares and the operation of machinery. Impulse noise consists of irregular spikes or pulses of relatively high amplitude and short duration.

Industrial, Scientific, and Medical (ISM) band   A range of frequencies where wireless communications can occur without first requiring operators to obtain a license.

Infrared   Very high frequencies just below visible light in the electromagnetic spectrum.

Infrastructure wireless LAN   The use of an access point by wireless devices communicating with each other and/or with wired stations connected to the access point.

Intersymbol interference A modulation rate that results in one symbol interfering with another.

Institute of Electrical and Electronic Engineers (IEEE) A private organization tasked by the American National Standards Institute with developing LAN standards to include wireless LAN standards.

ISDN Integrated Service Digital Network.

ISM Industrial, scientific and medical.

ISP Internet Service Provider.

ITU International Telecommunications Union.

Kilo Thousand.

LF Low Frequency.

Macro A prefix for millionth.

M-ary A term used to refer to the density of a constellation pattern as well as the number of phase shifts in a modulation scheme.

Mean Optimum Score (MOS) A subjective voice quality rating system that uses a scale of 1 to 5, with 5 representing the highest quality of voice.

Mega A prefix for million.

MF Medium Frequency.

MHz Megahertz.

Milli A prefix for thousand.

Modulation The process of changing the structure of a carrier signal to impress information on the signal.

MOS Mean Optimum Score.

Multipath distortion Reflections of a signal off objects that result in multiple signals, each slightly delayed due to different paths received at a receiver.

Nano   A prefix for billionth.

Network Address Translation (NAT)   The process of converting IP addresses used by stations behind a device to one or more different IP addresses for use on the other side of the device.

Non Return to Zero (NRZ)   a signaling technique under which binary 1s result in a digital signal staying at a high voltage while binary 0s are represented by a lack of voltage.

NII   National Information Infrastructure.

nm   Namometer.

NRZ   Nor Return to Zero.

Nyquist Relationship   The relationship between bandwidth (W) and the signaling rate in baud (B) on a channel such that B = 2W.

OFDM   Orthogonal Frequency Division Multiplexing.

Ohm's Law   The relationship between current (I), voltage (V) and resistance (R) such that V = I × R.

Orthogonal Frequency Division Multiplexing (ODM)   A transmission method under which multiple carriers are transmitted orthodonal or at right angles to each other over a wide range of frequencies.

Path loss   The signal attenuation between transmit and receive antennas.

PDA   Personal Digital Assistant.

Peer-to-peer networking   Wireless networking in which two devices communicate directly with each other.

Penta   A prefix for trillion.

Period   The time required for a signal to be transmitted over a period of one wavelength.

Phase shift modulation (PSM) A modulation technique under which the phase of a signal is shifted between two values to represent binary 0 and binary 1.

Phase modulation (PM) The process of varying a carrier signal with respect to the origination of its cycle.

Phase Shift Keying (PSK) Another term for single-bit phase modulation under which the phase of a carrier is altered between two values to modulate binary 0s and binary 1s.

PIFS Point Coordination Interframe Space

Ping A utility program that provides the round-trip delay to the target host. Ping represents a popular test program as a response indicates the pinging device has an operational network connection.

Point Coordination Interframe Space (PIFS) A time delay used by an access point to gain access to the medium.

Portal A public area such as a hotel conference room that supports wireless access to the Internet.

PPM Pulse Position Modulation.

PPPoE Point-to-Point Protocol over Ethernet.

Processing gain In a frequency hopping spread spectrum communications system the ratio of the bandwidth of each hop to the bandwidth of the transmission channel in dB.

Propagation loss The loss in signal strength of a signal as it traverses a medium.

PSK Phase Shift Keying.

PSTN Public Switched Telephone Network.

QAM Quadrature Amplitude Modulation.

Quadrature Amplitude Modulation (QAM) A modulation technique under which a group of bits are modulated using phase and amplitude changes.

Reset button   A button on some electronic products which, when pressed, returns the configuration of the device to factory default settings.

RF   Radio frequency.

RFC   Request For Comment.

Roaming   Moving from one location to another that results in service being provided by a different access point.

RTS   Request to Send.

Shannon's Law   A relationship between coding and noise on a channel and its effect on the maximum obtainable bit rate. The relationship developed by Professor Shannon is $C = W \log_2(1 + S)/N$, where C is the capacity in bits per second, W is the bandwidth in Hertz, S is the power of the transmitter and N is the power of thermal noise.

SFD   Start of Frame Delimiter.

Shared Wireless Access Protocol (SWAP)   The protocol that forms the basis for wireless home networking.

SHF   Super High Frequency.

Short Interframe Space (SIFS)   A time delay used to separate transmissions belonging to an immediate action, such as the transmission of an RTS, CTS, or ACK.

Signal constellation   A diagram that shows all possible signal points a device can transmit.

Signal-to-noise ratio (S/N)   The ratio of signal strength to noise power which is measured in dB and used to categorize the quality of transmission.

SIFS   Short Interframe Space.

Service Set identification (SSID)   A wireless network identification number.

Slow hop   A frequency hopping spread spectrum communications system in which the hopping rate is less than the frame rate.

Spreading code   A pseudo random binary string used to map a bit into a larger number of bits to provide redundancy. A spreading code permits a narrowband communications system to obtain wideband characteristics. Spreading codes are used in Direct Sequence Spread Spectrum transmission.

Spread spectrum communications   Communications where a narrowband signal is spread over a wider frequency band. Originally developed for military operations to compensate for enemy jamming, spread spectrum communications is used in commercial operations to include wireless LANs to minimize the effect of interference.

SSID   Service Set identification.

Stream cipher   A cipher which operates by expanding a short key into an infinite pseudorandom key stream.

Supplicant   A term used in the 802.1x standard to reference a client.

SWAP   Shared Wireless Access Protocol.

TDMA   Time Division Multiple Access.

Thermal noise   Noise generated by the movement of electrons or basic radiation from the sun. Thermal noise is also referred to as gaussian noise and represents a near uniform distribution of energy over the frequency spectrum.

Tribit encoding   The process of encoding three bits into a single signal change.

UHF   Ultra High Frequency.

ULF   Ultra Low Frequency.

UNII   Unlicensed National Information Infrastructure

USB   Universal Serial Bus.

VCS   Virtual Carrier Sense.

VHF   Very High Frequency.

Virtual Carrier Sense (VCS)   A derivative of the Carrier Sense Multiple Access with Collision Avoidance (CSMA/CD) access protocol under which a station first transmits a Request to Send (RTS) packet that must be acknowledged by a Clear to Send (CTS) packet from the receiver prior to transmission commencing.

VLF   Very Low Frequency.

Wavelength   The period of an oscillating signal.

WEP   Wired equivalent privacy.

Wired Equivalent privacy (WEP)   The security techniques used by IEEE 802.11 LANs.

Wireless bridge   A device designed to interconnect to wired LANs via wireless transmission. Most wireless bridges consist of an access point and a separate antenna that can be mounted on the side or top of a building.

INDEX

Note: Boldface numbers indicate illustrations.

access control, routers/gateways and, 213—214
access point-based wireless LANs, 282—283
access points, 6, 51, 52, 114—123, 282—284
   ad hoc wireless LANs, 121
   antenna connector on, 116
   back panel of, 116, **116**
   Basic Service Set (BSS) and, 118—122
   cost of, 283—284
   encryption settings for, 121
   equipment connection to, 115—116
   evolution of, 115
   Extended Service Set (ESS) and, 119—122
   features selection checklist for, 122
   IEEE 802.11 standard and, 150—151
   IEEE 802.11b standard and, 115
   infrastructure wireless LANs, 121
   multiple, network configuration for,
      118—122, **119**
   power source for, 115
   security issues for, 120—121
   single, network configuration for, 117—118, **117**
   SMC 2652 example of, 120, **120**
   Web browser to reset, 115—116
   Wired Equivalent Privacy (WEP) standard
      for, 120
ACK Frame, 178—179, **178**
active scanning, 179, **181**
ad hoc wireless LANs, access points and, 121
ad hoc mode, 179
adapter type network cards, 123
adaptive pulse code modulation (ADPCM), 248
Address fields, 175—176
addressing, 114, 175
   routers/gateways and, 136, 139, 204, 215—216,
      215
administration settings, network cards, 231—232
Administrator's Toolbox router option, 205,
   206, 207

Agere Systems Orinoco PC Card, 141—142, **142**,
   227—236, **228**
AiroPeek wireless protocol analyzer, 260—266
amplitude, 29, 64
   amplitude modulation (AM), 43, 64—65, **65**
   quadrature (QAM), 72—74, **73**, 82—84, **83**, 104,
      108—109, 193
antenna
   access points and connector for, 116
   diversity, 59, 135
   positioning, 58, 59, 198—199
Antheil, George, FHSS development and, 89
applications for wireless LANs, 11—16, 49—50,
   49
association, 180, **181**
asymmetrical digital subscriber line (ADSL), 105
asynchronous communications, Home RF
   standard and, 247—248
attenuation, 24, 52—56
authentication, 180, 258—260
   Extensible Authentication Protocol (EAP),
      258—260
   network cards and, 220—221

backoff algorithms, 167
backward compatibility issues, 256
bands of frequency, 46—48
bandwidth, 30—32
   coded OFDM (COFDM) and, 107, **108**
   DSSS and, 100—101, **101**
   Part 15 ruling and, 252—255
bandwidth spreading, DSSS and, 100—101, **101**
Barker code, 157, 184, **184**
Basic Service Set (BSS), 118—122, 148—149, 179
Basic Service set ID (BSSID), 175
battery life, 17, 18
baud, 24, 29, 42—43
bel, 32—34

Bell, Alexander G., 32
benefits of wireless LANs, 16
binary phase shift keying, 71, 81, 108—109, 193
    differential (DBPSK), 75—76, **76**
bit error rate (BER), 57
bit rate, 42
bits, 24, 42—43
bits per second, 42
Blowfish encryption algorithm, 243
Bluetooth, 31, 95, 254—255
bottlenecks, 17
BreezeNet DS.11 wireless bridge, 135, **136**
bridges, 127—136, **128**
    antenna diversity and, 135
    BreezeNet DS.11 wireless bridge, example of,
        135, **136**
    Ethernet frame example and, 129
    features selection checklist for, 133—136
    limitations of, 132
    operation of, 128
    overview of, 129
    port address table and, 129, 130—132, **130**
    rationale for use of, 127—128
    time stamps and, 132
    wireless operation using, 132, **133**
browsers, access point setting using, 115—116

carrier, 42
Carrier Sense Multiple Access with Collision
    Avoidance (CSMA/CA), 162, 240
Carrier Sense Multiple Access with Collision
    Detection (CSMA/CD), 163
carriers, 3
    phase modulation and, 67—68
    cellular phones, roaming, 181—182
    chip or chipping code, 10
    DSSS and, 97, 98—100
Clear to Send (CTS) packets, 164—165, **165**
client-based wireless LANs, economics of,
    280—281
Client Manager setting, network cards and,
    232—234, **233**
client support, IEEE 802.11 standard and,
    151—152

coded OFDM (COFDM), 59, 86, 104—111
    advantages of, 110
    bandwidth and, 107, **108**
    binary phase shift keying (BPSK) in, 108—109
    coded method in, 106—107
    coding in, 108—109
    data rate or transmission rate of, 109, 110
    disadvantages of, 110
    distortion vs., 110
    evolution of, 104—105
    fading vs., 110
    fast Fourier transform (FFT) and, 109
    forward error correction (FEC) in, 106
    frequency allocation and, 107
    frequency division multiplexing (FDM) in,
        105—106, **106**
    IEEE 802.11a and, 106—107, 111
    interference and, 107, **108**
    multipath interference vs., 110
    operation of, 107
    orthogonal frequency division multiplexing
        (OFDM) in, 106
    overview of, 105—107
    quadrature amplitude modulation (QAM) in,
        108—109
    quadrature phase shift keying (QPSK) in,
        108—109
    scrambling in, 108—109
    subcarrier use of, 110
college/campus use of wireless LANs, 13
collision avoidance, 167
collisions, 162—165
communications systems, 20, 85—111
complementary code keying (CCK), 78—79, 183,
    185—189
components of wireless LANs, 51—52, **51**
compression, data, 244
connection point (CP), Home RF standard and,
    240
constraints of wireless LANs, 17—18
Control Field, DSSS, 168—173
cost of wireless LANs, 17
coverage areas, 257
CRC field, 177

CTS Frame, 177—178, **178**
current, 35
cycle per second (CPS), 26

data compression, 244
Data Link Layer, 147—148
    Home RF standard and, 239
data rates, coded OFDM (COFDM) and, 109, 110
decibel, 34—35
    power measurement vs., 39, **40**
decibel above one milliwatt (decibel milliwatt),
        35—36
demodulation, 63
destination address (DA), 175
DHCP Server setup for, 209—210, **209**, 218
diagnostic settings, network cards and, 234, **234**
dibit encoding, 43, 69
differential binary phase shift keying (DBPSK),
        75—76, **76**, 157
differential quadrature phase shift keying(
        (DQPSK), 75, 76—78, **77**, 157, 187—189
diffuse wireless LANs, 9
Digital Audio Broadcasting (DAB), 105
Digital Enhanced Cordless Telephone (DECT),
        239, 241
direct sequence spread spectrum (DSSS), 10—11,
        20, 74—75, 86, 88, 89—91, **90**, 97—103, 196, 253
    advantages of, 101—102
    backward compatibility issues, 256
    bandwidth spreading in, 100—101, **101**
    Barker code for, 157, 184, **184**
    cell capacity in, 103
    chip or chipping code in, 97, 98—100
    complementary code keying (CCK) and, 183,
        185—189
    cost of, 102—103
    differential binary phase shift keying
        (DBPSK) in, 157
    differential quadrature phase shift keying
        (DQPSK) in, 157, 187—189
    disadvantages of, 102—103
    FHSS vs., 90—91, 102—103
    frame format in, 158—159, **159**
    frequency allocation in, 158
    Hadamard encoding in, 187
    IEEE 802.11 standard and, 98, 146, 157—159, 182
    IEEE 802.11b standard and, 146, 183
    industrial, scientific, medical (ISM) band for, 98
    modulation and, 157, 183—189
    multipath tolerance in, 189
    noise and, 97
    operation of, 98—99
    overview of, 157
    power usage of, 102—103
    privacy in, 102
    processing gain (PG) in, 101
    pseudonoise (PN) code in, 97
    regulations concerning, 97—98
    Shannon's law and, 100—101
    signal to noise (S/N) ratio in, 101
    spreading codes, 186
discrete multitone (DMT), 105
distortion, coded OFDM (COFDM) vs., 110
Distributed Inter Frame Space (DIFS), 163
distribution systems, 149, **150**
diversity, 59
    antenna, 135
    space, 135
Domain Name Servers (DNS), 204
drivers, network card, 219—222
Duration/ID field, 175
dwell time, 10, 92, 93
Dynamic Host Configuration Protocol
        (DHCP), 136, 205, 209, 218, 280—281

eBay auction site, for hardware, 271—278
Echo packet, 203
economics of wireless LANs, 279—284
educational applications for wireless LANs, 13
electro optical frequency (EOF), 47
electromagnetic interference (EMF), 18
encoding
    coded OFDM (COFDM) and, 108—109
    dibit, 69
    Gray code, 82—84, **82**
    Hadamard, 187
    nonreturn to zero (NRZ), 64
    tribit, 69

encryption, 172—175
   Blowfish, 243
   shared key, 220—221
   Wired Equivalent Privacy (WEP), 172—175
encryption settings, for access points, 121
enhancing reception, 58—59
ESSID, 150
Ethernet, 17, 63
   bridges and, 129
   point to point protocol over Ethernet
      (PPPoE) and, 142
ETSI Digital Video Broadcasting Terrestrial
   (DVB—T), 105
European Telecommunications Standard
   Institute (ETSI), 105
extended interframe space (EIFS), 166
Extended ISA (EISA) network cards, 123
Extended Service Set (ESS), 119—122, 148, 149
Extensible Authentication Protocol (EAP),
   258—260
extension points, 6—8, **8**
extremely high frequency (EHF), 47, 48
extremely low frequency (ELF), 47, 48

fading, 57—58
   coded OFDM (COFDM) vs., 110
Fast Ethernet, 17
fast Fourier transform (FFT), 109
fast hop or fast frequency hopping system,
   FHSS, 94
Federal Communications Commission (FCC), 3,
   30, 46
   DSSS and, 97—98
   FHSS and 91—93
   Home RF standard and, 238
   Part 15 ruling and, 252—255
fiber optics, 31
filtering, 213—214
firewalls, 114
forward error correction (FEC)
   coded OFDM (COFDM) and, 106
   FHSS and, 95
Frame body field, 176
frame formats, 167—177, **168**

ACK Frame, 178—179, **178**
CTS Frame, 177—178, **178**
DSSS and, 158—159, **159**
FHSS and, 155—156, **155**
Home RF standard and, 246—247
IEEE 802.11 standard and, 167—177, **168**
IEEE 802.11a standard and, 191—192, **191**
infrared (IR) and, 161—162, **160**
MAC Data Frame, 168—177
RTS Frame, 177, **177**
superframes, 247
free space, ideal, 53
frequency, 25, 26—27, 29, 64
frequency allocation, 3, 21, 24, 30, 45—50
   band nomenclature in, 46—48, **47, 48**
   coded OFDM (COFDM) and, 107
   DSSS and, 158
   IEEE 802.11 standard and, 153
   IEEE 802.11a standard and, 190
   Part 15 ruling and, 252—255
   United States, 46
   wireless LAN, 50
frequency division multiplexing (FDM),
   105—106, **106**, 190
frequency hopping (*See* hopping)
frequency hopping spread spectrum (FHSS),
   10—11, 20, 79, 86, 88—89, **88**, 91—97, **93**
   advantages of, 95—97
   band comparison for, **93**
   cell capacity in, 103, **103**
   DSSS vs., 90—91, 102—103
   dwell time in, 92, **93**
   fast hop or fast frequency hopping system
      in, 94
   forward error correction (FEC) in, 95
   frame format, 155—156, **155**
   frequency allocation and, 153
   geographic overlap of systems in, 96, **97**
   Home RF standard and, 241—242
   hopping channels and, 153—154
   hopping modes in, 94—95
   hopping rate in, 93
   hopping sequence and, 156
   IEEE 802.11 and, 92, 146, 153

modulation and, 154—155
multipath interference vs., 96, **97**
noise and, 95—97, **96**
operational parameters of, 93
packet transmission capability of, 94
Part 15 ruling and, 252—255
processing gain (PGA) in, 95—96
regulations concerning, 91—93
slow hop or slow frequency hopping system
  in, 94—95
frequency modulation (FM), 64, 65—67, **66**
frequency shift keying (FSK), 42, 66
  Gaussian shaped, 154—155
frequency shift modulation, 66
frequency spectrum, 24, 29—31, 30
FromDS subfield, 169
future developments, 22, 251—266

gateways (*See also* routers and gateways), 114,
  136—143
Gaussian filter, 80
Gaussian frequency shift keying (GFSK), 79—80
Gaussian noise, 37—38, **37**, 80
Gaussian shaped FSK (GFSK), 154—155
giga-, 25
Gray code, 82—84, **82**

Hadamard encoding, 187
handoffs, 6
handshaking, 104—105
hardware, 20, 21, 113—143
  access points (*See* access points)
  bridges, 127—136, **128**
  eBay auction as source of, 271—278
  manufacturers of, listing for, 267—278
  network cards for, 123—127, 218—226
  router/gateway, 136—143
Hertz, 26
high energy frequency (HEF), 47
high frequency (HF), 47, 48
Home RF standard, 31, 43, 237—249
  adaptive pulse code modulation (ADPCM) in,
    248
  asynchronous support in, 247—248

connection point for, 240
data compression in, 244
Data Link Layer specifications for, 239
device support for, 242—243
Digital Enhanced Cordless Telephone (DECT)
  and, 239, 241
frame duration and type in, 246—247
frequency hopping spread spectrum (FHSS)
  and, 241—242
IEEE 802.11 and, 238
interference and, 254—255
Media Access Control (MAC) layer and,
  245—246, **246**
modulation in, 242
network architecture of, 240
nodes for, 240
operations of, 244—249
Physical Layer specifications for, 239, 245
power use in, 242
Public Switched Telephone Network (PSTN)
  and, 241
security for, 243—244
Shared Wireless Access Protocol (SWAP) and,
  238, 241, 254—255
superframes in, 247
system requirements for, 241
technical characteristics of, 241
Time Division multiple access (TDMA) in,
  240, 245
versions of, 239
voice transport in 248—249
Wireless Home Networking in, 238
hopping, 3, 10—11, 58
hopping channels, FHSS, 153—154
hopping modes, FHSS, 94—95
hopping rate, FHSS, 93
hopping sequence, FHSS, 156
hospital use of wireless LANs, 12—13

ideal free space, 53
IEEE 802.11 standard, 21, 146—156
  access points and, 150—151
  authentication and association under, 180, **181**
  basic service set (BSS) in, 148—149

Carrier Sense Multiple Access with Collision Avoidance (CSMA/CA) in, 162, 240
Carrier Sense Multiple Access with Collision Detection (CSMA/CD), 163
Clear to Send (CTS) packets, 164—165, **165**
client support in, 151—152
collision avoidance in, 167
collisions and, 162—165
Data Link Layer in, 147—148
direct sequence spread spectrum (DSSS) and, 157—159, 182
Distributed Inter Frame Space (DIFS) in, 163
distribution system in, 149, **150**
DSSS and, 98
extended service set (ESS) in, 148, 149
extensions to, 182—189
FHSS and, 92
frame format in, 155—156, **155**, 158—159, **159**, 161—162, **160**, 167—177
frequency allocation and, 153, 158
frequency hopping spread spectrum (FHSS) and, 153
Home RF standard and, 238
hopping channels and, 153—154
hopping sequence and, 156
infrared (IR) and, 159—161
interframe spaces and, 166
joining an existing cell under, 179—180
layers of, 147
Logical Link Control (LLC) in, 147
MAC Service Data Units (MSDUs) in, 147, 148
Media Access Control (MAC) layer and, 162—179
modulation and, 154—155, 157, 160—161
network allocation vector (NAV) in, 164
operations using, 179—182
overview of, 147—148
packet transmission in, 163—165
Physical Layer and, 148, 152—154
Physical Layer Convergence Procedure (PLCP) sublayer in, 148
physical medium dependent (PMD) sublayer in, 148
portals and, 152

radio frequency (RF) in, 148
Request to Send (RTS) packets and, 163—165, **164**
roaming under, 180—182
send-and-wait algorithm in, 171
topologies and, 148—152, 148
transmission distance/range, 152
transmission rates and, 182
virtual carrier sense (VCS) in, 163—165, 166
IEEE 802.11a standard, 21, 84, 190—193
coded OFDM (COFDM) and, 106—107, 111
frame format, 191—192, **191**
frequency allocation and, 190
modulation, 191
operations under, 193
IEEE 802.11b standard, 21, 78—79, 146, 183—189
access points and, 115
complementary code keying (CCK) and, 183, 185—189
direct sequence spread spectrum (DSSS) and, 183
modulation and, 183—189
operation under, 183
Wireless Equivalent Privacy (WEP) and, 183
IEEE 802.11g standard, 255—257
IEEE 802.11x standard, 257—260
IEEE wireless LAN standards, 21
impairments to wireless LANs, 50—59
impulse noise, 38, **38**
industrial, scientific, and medical (ISM) band, 9, 21, 50, 30—31, 98, 253—254
Industrial Standard Architecture (ISA) network cards, 123
infrared (IR), 8—9, 159—161
frame formats in, 161—162, **160**
IEEE 802.11 standard and, 146, 159—161
modulation in, 160—161
on-off keying (OOK) in, 160
pulse position modulation (PPM) in, 160
infrastructure wireless LANs, access points and, 121
installation, 195—236
Agere Systems Orinoco PC Card, 141—142, **142**, 227—236, **228**

Home RF standard, 22
network cards, 218—226
router, SMC Networks barrier router, 196—218
testing of, 234—236, **235**
Institute of Electrical and Electronics Engineering (IEEE), 21, 146
Integrated Service Digital Network (ISDN), 63
interference, 17, 18, 24, 31, 58
coded OFDM (COFDM) and, 107, **108**
FHSS and, 96, **97**
intersymbol, 43
radio frequency, 254—255
interframe spaces, 166
International Telecommunications Union (ITU), 45
Internet access, 14, 114
Internet Connection Sharing, 280—281
Internet portals, 152
Internet Service Providers, 204
intersymbol interference, 43
inventory control via wireless LANs, 13—14
IP addresses, 136, 139, 204
IRE Transactions Information Theory, 185

jamming, FHSS and vs., 87
joining an existing cell, 179—180

keying, 66
kilo-, 25

Lamar, Hedy, FHSS development and, 89
layers of IEEE 802.11 standards, 147, **147**
Lempel-Ziv-Ross Williams compression, 244
light, speed of, 28
line of sight, 9
link quality setting, network card, 223—225, **224, 225**
Logical Link Control (LLC), 147
low frequency (LF), 47, 48
LZRW3-A data compression algorithm, 244

M-ary operation, 71—72
MAC Data Frame, 168—177

Address fields in, 175—176
Control field of, 168—173
CRC field in, 177
Duration/ID field in, 175
Frame body field in, 176
FromDS subfield of, 169
More data subfield in, 172
More Fragments subfields of, 169—170
Order subfield of, 175
Power Management subfield in, 171—172
Protocol Version subfield of, 169
Retry subfield in, 171
send-and-wait algorithm, 171
Sequence control field in, 176
TODS subfield of, 169
Type and Subtype subfields of, 169, 170
WEP subfield in, 172—175
MAC Service Data Units (MSDUs), 147, 148
mapping, virtual server, 211
Media Access Control (MAC) layer, 147, 162—179
Carrier Sense Multiple Access with Collision Avoidance (CSMA/CA) in, 162, 240
Carrier Sense Multiple Access with Collision Detection (CSMA/CD), 163
Clear to Send (CTS) packets, 164—165, **165**
collision avoidance in, 167
collisions and, 162—165
Distributed Coordination Function (DCF), 163
Distributed Inter Frame Space (DIFS) in, 163
Home RF standard and, 245—246, **246**
interframe spaces and, 166
network allocation vector (NAV) in, 164
packet transmission in, 163—165
Request to Send (RTS) packets and, 163—165, **164**
virtual carrier sense (VCS) in, 163—165, 166
medium frequency (MF), 47, 48
mega-, 25
micro-, 25
microwave communications, 9
microwave ovens, 31
milli-, 25

mobile nodes, 51, 52, 149
  Home RF standard and, 240
modems, 105, 114
modulation, 3, 19—20, 42—43, 58, 61—84
  adaptive pulse code modulation (ADPCM),
    248
  amplitude, 64—65, **65**
  Barker code for, 184, **184**
  binary phase shift keying (BPSK) and, 71, 81,
    108—109, 193
  combined, 72—74
  complementary code keying (CCK) and,
    78—79, 183, 185—189
  differential binary phase shift keying
    (DBPSK), 75—76, **76**, 157
  differential quadrature phase shift keying
    (DQPSK) in, 75, 76—78, **77**, 157, 187—189
  direct sequence spread spectrum (DSSS),
    74—75, 157
  discrete multitone (DMT), 105
  frequency, 64, 65—67, **66**
  frequency hopping spread spectrum (FHSS),
    79
  frequency shift, 66
  Gaussian frequency shift keying (GFSK),
    79—80
  Gaussian shaped FSK (GFSK), 154—155
  Home RF standard and, 242
  IEEE 802.11 standard and, 154—155
  IEEE 802.11a standard and, 191
  IEEE 802.11b standard and, 183—189
  infrared (IR) and, 160—161
  M-ary operation and, 71—72
  maximum rate of, 43—44
  multilevel phase shift keying and, 69—70, **68**
  on off keying (OOK) in, 160
  orthogonal frequency division multiplexing
    (OFDM), 79, 80—82, **81**
  phase, 64, 67—74
  phase shift, 68—69
  process of, 63—64
  pulse position modulation (PPM) in, 160
  quadrature amplitude (QAM), 72—74, **73**,
    82—84, **83**, 104, 108—109, 193

quadrature phase shift keying (QPSK) and,
    71—72, **72**, 81, 108—109, 193
  radio frequency (RF), 63—64
  rationale for, 62—63
  sine waves and, 63—64, 67
More data subfield, 172
More Fragments subfields, 169—170
Morse code, 66
multicarrier transmission, 105
multilevel phase shift keying, 69—70, **68**
multipath propagation/interference/distortion,
    18, 24, 39, 56—58, **56**
  coded OFDM (COFDM) and vs., 110
  FHSS and, 96, **97**
multipath tolerance, 189
multiple access point network, 118—122, **119**
multiplexing, 59
  frequency division multiplexing (FDM) in,
    105—106, **106**
  orthogonal frequency division multiplexing
    (OFDM) in, 106
multitone transmission, 59, 105

nano-, 25
National Information Infrastructure (NII), 50
National Telecommunications and
    Information Administration (NTIA), 46
network address translation (NAT), 136,
    138—140, 212
network allocation vector (NAV), 164
network cards, 123—127, 218—226
  adapter configuration for, 220
  adapter type, 123
  administration settings for, 231—232
  Agere Systems Orinoco PC Card, 141—142,
    142, 227—236, **228**
  authentication in, 220—221
  channel property for, 221
  Client Manager setting for, 232—234, **233**
  configuration utility for, 222—226, **223**
  diagnostic settings for, 234, **234**
  driver installation for, 219—222
  Extended ISA (EISA) type, 123
  features selection checklist for, 126

Industrial Standard Architecture (ISA) type, 123
link quality setting for, 223—225, **224, 225**
network type property for, 221
PCI type, 123—125, **124**
power save mode for, 221
profile options for, 228—230, **229, 230**
RTS threshold setting for, 221
Service Set ID (SSID) for, 221—222
SMC Networks PCI type card, example of,
    123—125, **124**
SMC Networks Wireless PC card, example
    of, 125—127, **126**
testing installation of, 234—236, **235**
types of, 123
USB type, 123
Wireless Equivalent Privacy (WEP) and, 222,
    225—226
network configuration, 4—6, **5**
network ID, routers/gateways, 216—217
networking via wireless LANs, 15
nodes (*See also* mobile nodes), Home RF
    standard and, 240
noise, 37—40, 80
    DSSS and, 97
    FHSS and, 95—97, **96**
nonreturn to zero (NRZ) encoding, 64
Nyquist Relationship, 24, 41—45, 68

office wireless LANs, 51—52, **51**
Ohm's law, 35
on—off keying (OOK), 160
open systems, 221
operation of wireless LANs, 304, 3
Order subfield, 175
Orinoco Residential Gateway, 141—142, **142**,
    227—236, **228**
orthogonal frequency division multiplexing
    (OFDM), 10—11, 20, 59, 79—82, **81**, 86, 106,
    190
    backward compatibility issues, 256
    coded (*See* coded OFDM)
    frame format in, 191—192, **191**
    IEEE 802.11g standard and, 256
    modulation, 191

packet filtering in, 213—214
packet transmission, 94, 163—165
Pad field, 192
Part 15 ruling and, 252—255
passive scanning, 180
path loss, 24, 41, 52—56, **55**
Payload Field, 192
PCI type network card, example of, 123—125,
    124
peer-to-peer networks, 4, **5**, 149, **149**
penta-, 25
period of signal, 26—27
Personal Communication Systems (PCS), 41
phase, 29, 43, 64
phase modulation, 64, 67—74
    binary phase shift keying (BPSK) and, 71, 81,
        108—109, 193
    M-ary operation and, 71—72
    multilevel phase shift keying and, 69—70, **68**
    phase shift, 68—69
    quadrature amplitude modulation (QAM)
        and, 72—74, **73**
    quadrature phase shift keying (QPSK) and,
        71—72, **72**, 81, 108—109, 193
    sine wave in, 67
    single bit, 68—69
    varying the carrier in, 67—68
phase shift keying (PSK)
    binary, 71, 81, 108—109, 193
    complementary code keying (CCK) QPSK,
        78—79
    differential binary (DBPSK), 75—76, **76**, 157
    differential quadrature (DQPSK) in, 75,
        76—78, **77**, 157, 187—189
    multilevel, 69—70, **68**
    quadrature, 71—72, **72**, 81, 108—109, 193
phase shift modulation, 68—69
Physical Layer, 147—148
    Home RF standard and, 239, 245
    IEEE 802.11 standard and, 152—154
Physical Layer Convergence Procedure (PLCP),
    148, 192
Physical Medium Dependent (PMD) sublayer,
    148

ping, 203
PLCP field, 192
point coordination interframe space (PISF), 166
point-to-point protocol over Ethernet (PPPoE), 142
polyphase complementary code, 186
port address table, 129. 130—132, **130**
portals, Internet, 14, 152
ports, routers/gateways and, 196—197
power
    decibel vs., 39, **40**
    DSSS and usage, 102—103
    loss of, 52—56, **55**
Power Management subfield, 171—172
power measurement, 32—36
power ratios, 53—56
power save mode, network card, 221
power source, for access points, 115
powers of ten, 25
Primary Setup router configuration screen, 207—209, **207, 208**
privacy issues
    DSSS and, 102
    Wired Equivalent Privacy (WEP) standard for, 120, 172—175, 183, 222, 225—226, 243—244, 257—257
processing gain (PG)
    DSSS and, 101
    FHSS and, 95—96
profile options, network card, 228—230, **229, 230**
propagation delay, 24
propagation loss, 40—41, 52—56
protocol analyzer, wireless (AiroPeek), 260—266
protocol version subfield, 169
pseudonoise (PN) code, DSSS and, 97
Public Switched Telephone Network (PSTN), 62—63, 69, 241
pulse code modulation, adaptive (ADPCM), 248
pulse position modulation (PPM), 160
pulse shaping, 80

quadrature amplitude modulation (QAM), 72—74, **73**, 82—84, **83**, 104, 108—109, 193

quadrature phase shift keying, 71—72, **72**, 81, 108—109, 193
    complementary code keying (CCK) QPSK, 78—79
    differential ((DQPSK), 75, 76—78, **77**, 157, 187—189

radio frequency (RF), 9—11, 3, 46, 148
modulation of, 63
range of transmission, 17
    IEEE 802.11 standard and, 152
reception enhancement, 58—59
recipient address (RA), 175
reflection, multipath propagation and, 56—57, **56**, 58
reflective wireless LANs, 9
repeaters, 6—8
Request to Send (RTS) packets, 163—165, **164**
resistance, 35
retry subfield, 171
RF interference, 254—255
    roaming, 6, 7, 6, 180—182
    cellular vs. wireless, 181—182
router-based wireless LANs, 282—283
routers and gateways, 114, 136—143
    access control using, 213—214
    addressing in, 136, 139, 204
    addressing options for, 215—216, **215**
    Administrator's Toolbox option in, 205, **206,** 207
    antenna positioning for, 198—199
    configuration of, 204—205
    configuration options for, 205—216
    connectivity tradeoffs in installation of, 199, **200**
    DHCP Server setup for, 209—210, **209**, 218
    Dynamic Host Configuration Protocol (DHCP) and, 136, 205, 209, 218
    features selection checklist for, 142—143
    installation of, 196—218
    network address translation (NAT) in, 136, 138—139, 140, 212
    network configuration using, 137—138, **138**
    network ID for, 216—217, **217**

Orinoco Residential Gateway, example of, 141—142, **142**
overview of, 137—138
packet filtering in, 213—214
point to point protocol over Ethernet (PPPoE) and, 142
ports on, 196—197
Primary Setup screen for, 207—209, **207, 208**
site location for, 198
SMC Barricade, example of, 139—141, **140**
software setup for, 202
verifying computer—router connectivity in, 203—204
virtual private network (VPN) support in, 197
Virtual Server setup for, 210—211, **211**
WINIPCFG configuration of, 200—202, 217—218
wireless settings for, 216—217, **217**
RTS Frame, 177, **177**

satellite, 46
scanning, active and passive, 179—180, **181**
Schneier, Bruce, 243
scrambling, coded OFDM (COFDM) and, 108—109
second generation wireless, 31
security, 285—287
access control using router in, 213—214
access points and, 120—121
DSSS and, 102
encryption settings for, 121
Extensible Authentication Protocol (EAP), 258—260
Home RF standard and, 243—244
packet filtering in, 213—214
Wired Equivalent Privacy (WEP) standard for, 120, 172—175, 183, 222, 225—226, 243—244, 257—257
send-and-wait algorithm, 171
Sequence control field, 176
Service field, 192
Service Set ID (SSID), network card, 221—222
Shannon's law, 24, 41, 44—45, 100—101
shared-key encryption, 220—221

Shared Wireless Access Protocol (SWAP), 238, 241, 254—255
shielding, 39
shift, frequency, 57
shifting, 79
short interframe space (SIFS), 166
signal constellation point diagram, 71—72, **71**, 74
signal-to-noise (S/N) ratio, 37—40
DSSS and, 101
sine waves, 26—27, **26**, 63—64
phase modulation and, 67
single access point network, 117—118, **117**
single-bit phase modulation, 68—69
slow hop or slow frequency hopping system, FHSS and, 94—95
SMC 2652 wireless access point, 120, **120**
SMC Networks
Barricade wireless router, 139—141, **140**, 196—218
EZ Connect PC Card, 218—226
PCI type card, example of, 123—125, **124**
Wireless PC card, example of, 125—127, **126**
space diversity, 135
spectrum (*See* frequency spectrum)
spread spectrum technology, 9, 86—91
development of, 86—87
direct sequence (*See* direct sequence spread spectrum)
frequency hopping (*See* frequency hopping spread spectrum)
general operation of, 87
jamming vs., 87
methods for, 88—89
spreading codes, 10, 89, 186
spreading ratio, 10
standards (*See also* IEEE 802.11xx standards), 21, 145—193
stations, 149
stream ciphers, 172—175
subcarriers, 11
coded OFDM (COFDM) and, 110
super high frequency (SHF), 47, 48
superframes, 247

TCP/IP, 114
Telebit Trailblazer, 104
terminology, 19, 23—58
testing network installation, 234—236, **235**
thermal noise, 37—38
throughput rates, 17—18
Time Division multiple access (TDMA), 240, 245
time stamps, bridges and, 132
TODS subfield, 169
Token Ring, 63
topologies, 20, 148—152
trade shows and wireless LANs, 15—16
training centers using wireless LANs, 14
transceiver, 52
transmission distance (*See* range of
    transmission)
transmission rates, 24, 29, 41—45
    coded OFDM (COFDM) and, 109, 110
    IEEE 802.11 standard, 146, 182
    IEEE 802.11a standard, 146
    IEEE 802.11b standard, 146
transmitter/receiver, 51
tribit encoding, 69
Type and Subtype subfields, 169, 170

U.S. Commerce Department, 46
ultra high frequency (UHF), 47, 48
ultra low frequency (ULF), 47, 48
unlicensed ISM band, 30—31

Unlicensed National Infrastructure (UNII)
    band, 50, 190
USB network cards, 123

very high frequency (VHF), 47, 48
very low frequency (VLF), 47, 48
virtual carrier sense (VCS), 163—165, 166
virtual private network (VPN) support,
    routers/gateways and, 197
Virtual Server router setup, 210—211, **211**
virtual servers, 197
    mapping, 211
voice transport, Home RF standard and
    248—249
voltage, 35, 64

watts, 35—36
wavelength, 25, 27—29
Web browser (*See* browsers)
WEP subfield, 172—175
white noise, 37—38, **37**
Williams, Ross, 244
WINIPCFG router configuration, 200—202,
    217—218
Wired Equivalent Privacy (WEP) standard for,
    120, 172—175, 183, 222, 225—226, 243—244,
    257—257
Wireless Home Networking, 238
wireless-to-wired networks, 4—6, **5**

ABOUT THE AUTHOR

GILBERT HELD is an internationally known author and lecturer in the field of computer communications. He has represented the United States at technical conferences in Jerusalem and Moscow and has taught seminars covering LAN performance and other communications topics literally around the globe. Gil is the only person to twice win the prestigious Karp award for technical excellence in writing. He has also won awards from Federal Computer Week and the Association of American Publishers. In his spare time Gil restores Corvettes.

www.ingramcontent.com/pod-product-compliance
Lightning Source LLC
Chambersburg PA
CBHW080351060326
40689CB00019B/3965